Environmental Sociology

Environmental Sociology

Risk and Sustainability in Modernity

Cristiano Luis Lenzi

LEXINGTON BOOKS
Lanham • Boulder • New York • London

Published by Lexington Books
An imprint of The Rowman & Littlefield Publishing Group, Inc.
4501 Forbes Boulevard, Suite 200, Lanham, Maryland 20706
www.rowman.com

6 Tinworth Street, London SE11 5AL, United Kingdom

British Library Cataloguing in Publication Information Available

Library of Congress Cataloging-in-Publication Data Available

ISBN 978-1-66691-150-3 (cloth)
ISBN 978-1-66691-151-0 (electronic)
ISBN 978-1-66691-152-7 (pbk.)

To my mother, Irma P. Lenzi, and Dayane, Nino e Mia.

Contents

Introduction

The meeting of sociology and ecology has been marked by a series of controversies. In recent years, several authors have signaled the need for sociologists to dedicate further attention to environmental issues, as such concerns we are presently facing about the environment must have roots in social processes. At the same time, those who defend such a reorientation in sociology recognize that such a task is not a simple one, as it advocates the preponderance of the "social" in relation to the "natural" that would rely on the sense of being of its own sociology.

The discussion started at the end of the 1970s and beginning of the 1980s, when American sociologists Riley E. Dunlap and William R. Catton Jr. proposed the creation of an environmental sociology. The starting point for the proposal made by those authors was justly a criticism of the emphasis of sociology on the social as opposed to a natural focus. They also argued that the sociological neglect of ecological factors would be welcomed in the range of the social sciences, as it would represent a positive change increasing marginalization of reductionist environmental explanations and providing value to specific sociocultural explanations. Although Catton and Dunlap did not intend to promote a return to the environmental reductionism that permeated sociological thinking in the nineteenth and twentieth centuries, they had doubts about the success the cultural paradigm of the social sciences would have achieved.

Particularly, there are two contradictory aspects that deserve consideration in relation to the impact of the work by Catton and Dunlap. On one side, enduring acceptance by ecological critics that both authors address regarding their condemnation of sociological negligence in relation to environmental issues is shared by several social scientists who address the theme, as will be seen in this book. On the other side, although there have been repercussions from the controversy generated by Catton and Dunlap, its theoretical penetration occurred in a partial way. This is because, although Catton and Dunlap have left their mark on the emergence and development of the debate about the possibility of creating an environmental sociology, much of what has come to be produced in recent years in the field of social sciences has

1

ended up not following the sociological direction that they propsed at the end of the 1970s. In the 1990s, they themselves recognized that their proposals failed (Martell 1994, 9).

But if these authors' environmental sociology was not accepted at first, in the 1990s some attempts appeared to re-create it or at least gave it new direction. Furthermore, in the 1990s the ecological issue was finally taken to the mainstream core of contemporary social theory.[1] Such new directions in environmental sociology can be seen in the impact the concepts of sustainable development (SD) and ecological modernization (EM) produced, along with the works by Anthony Giddens and Ulrich Beck, in the field of sociology in recent years.[2] For example, for authors such as Barry (1999a), the construction of a social ecological theory should involve the process of incorporating and enhancing some of the ideas contained in the SD discourse.. Others argue that the EM theory will be one of the theoretical currents that contribute the most to "greening" the social sciences (Mol, Spaargaren, and Buttel 2000). Such assessment is also addressed to the works performed by Giddens and Beck. For Goldblatt (1996), both have become references in the social sciences to understand the social and political dimensions of modern environmental degradation. And such opinion is confirmed by Buttel, for whom EM and the works written by Giddens and Beck represent the possibility to "correct deficiencies in American environmental sociology" (2000a, 34).

This study will provide a comparative analysis of those three approaches, with the aim to rethink environmental sociology as a new area of research within the contemporary social sciences. As I will discuss, there are reasons to believe that all of these approaches may serve as directives for such new areas in sociology research, addressing it to the fundamental issues concerning the relationship between society and the environment. I do not mean to say with this that those three perspectives exhaust the interest of such an area of study, but instead that they outline some issues that, according to my view, a sociology focused on environmental issues cannot avoid.

Some problems that arise before such a picture must be quickly considered. In most of the literature, SD is considered an ecological and political discourse instead of a sociological theory. This does not mean that it is not related to the sociological theorization per se, as it tends to be seen as a fundamental discourse to the greening of social theory. The same partially happens with the concept of EM. For authors like Dryzek (1997), both the latter and the former would constitute two different kinds of ecological discourses. But if the idea of EM is perceived in such a way by some social scientists, others have seen it as a sociological theory on its own terms.

Such approaches are frequently based on opposite or conflicting perspectives. Supporters of EM express criticism in relation to the concept of SD, considering their analytical purposes unsatisfactory, mainly because there is not only one but countless SD concepts. This was something that, according to them, would prevent the foundation of an environmental sociology. On the other hand, the notion of EM is criticized by supporters of SD issues—issues which would be emphasized precisely by the discussion of SD. Concurrently, the EM perspective is seen as being opposed to the theory of risk society (RS) developed by authors such as Beck (1992).

Disagreement and tension between these perspectives is not new if we look into the history of sociology, and more recently into the context in which environmental sociology has emerged. The criticism directed at sociology concerning its inability to accumulate knowledge is well known, since the diversity of languages and theories existing within it was seen in the past as a sign of intellectual immaturity. As will be seen in the first chapter, environmental sociology seems to have inherited such a reprehensible feature from sociology. Thus, the environmental issue has transcended its own field of sociology and has encompassed several areas of the social sciences (economy, history, geography, anthropology, political theory, etc.). This may seem to be a setback for those who have seen the possibility of a holistic approach in ecology and the disruption of any kind of specialization or fragmentation of knowledge that follows.

Upon such a landscape, there are two possible options. On one side, the positivist affirmation could be chosen that sociology would remain in a state of "scientific immaturity." Such a vision implies that the maturing of the discipline would occur after such disagreements dissipate. The second option (validated by this book) is to deal with such objections in a different way. The first option would perhaps have received greater support in the past, when the social sciences were thought to be following the same path as the exact sciences. But today, such a vision is far from being hegemonic. The interpretation that is seen in the objection and in the theoretical diversity of the endemic aspects of the social sciences has been widely accepted, and those phenomena are disconnected from the idea of "immaturity," as it was argued in the past.[3] Presently, it is recognized that diversity is not just a sign of creativity of the social sciences but also a result of the complexity of the issues involved in the study of human societies. Added to this, many differences between theories and authors are more apparent than real, or, looking through another lens, their differences are overvalued in relation to their common points. For instance, some of the main names in contemporary social sciences (Bourdieu, Habermas, Giddens) have been pointed out as representatives of a new theoretical movement marked by their search for a synthesis, thus implying that those authors constructed their work by approaching or even integrating great classics (Durkheim, Weber, Marx) and theories (structuralism, functionalism, symbolic interactionism) of sociology in a way that perhaps no one dared to perform previously. On the other hand, contemporary authors show that behind some apparently quite different theories there is a "socially unrecognized accumulation." For the most part, social theories share common points that simply are not admitted by their authors. There is a trend toward pointing out existing differences rather than similarities between these theories. This means that the difference or similarity between them cannot always be confirmed by facts and empirical data. Often such divergences are located in the set of suppositions contained in the theoretical discussion and are not always made explicit by their representatives. The difference, similarity, complementarity, or existing conflict between theories is not something given or simply perceived. The lines that divide disciplines, theories, and concepts can themselves be objects of debate and analysis.[4]

The present work is far from exhausting such discussion. It intends to demonstrate how the meaning of the greening of sociology is still controversial and to clarify a little the field of environmental sociology that at the same time proposes a different interpretation of some previously mentioned issues and perspectives. Our argument is that EM, SD, and RS can be seen as complementary perspectives for several points, and due to this, they respond to distinct interests of environmental sociology. Their differences do not necessarily mean incompatibility. One of the problems involving assessments of such perspectives is precisely the tendency to place them as a conflicting theoretical guidance or consider their differences as a reflection of incompatible aspects of environmental sociology. For these reasons, the efforts already made still seem timid to us in order to clarify how such approaches are interlaced and what consequences this would cause to each of them and to environmental sociology itself.

In general, this work points out a conceptual reconstruction of these three approaches in the field of environmental sociology, trying to make apparent the vital contribution they bring to socioenvironmental studies. Nevertheless, it will not provide what many would desire: a conceptual alternative structure that incorporates all these perspectives. I tried to investigate the possibility of converging the approach without necessarily diluting one into another. Thus, a more immediate consequence of this study suggests that it is possible to insert each of these approaches into one same work without accusing the researcher of an irresponsible eclecticism. By putting all those three perspectives side by side, we prevent relapse into a total and maybe unfair criticism of each of them. The concerns that afflict the supporters of a given approach may be enlightened by a concurrent approach without making them mutually exclusive. And in daring to analyze these three perspectives as a set, I hope to open the doors to the emergence of a new vision, and maybe to a new path of environmental sociology.

NOTES

1. These arguments can be found in Buttel (1987, 1996), Cohen (2000a), Goldblatt (1996), and Hannigan (1995).

2. In the remainder of this book, we will use the abbreviations EM for ecological modernization, SD for sustainable development, and RS for risk society.

3. For a discussion, see Alexander (1987), Bryant (1995), and Stehr (1982).

4. For a discussion of these aspects, see Alexander (1987) and Collins (1999).

Chapter 1

Greening Sociology

The Challenge of Environmental Sociology

The aim of this chapter is to perform a preliminary assessment on the relationship between sociology and the environmental issue. In the first part, we shall analyze the proposal for the creation of an environmental sociology made by Catton and Dunlap by the end of the 1970s. Catton and Dunlap had a fundamental role in starting the debate about the possibility of creating an environmental sociology. Their importance relies on the fact that they are among the first authors who tried to specifically define and identify environmental sociology. Secondly, we shall assess the new directions the sociological literature addressing the environmental theme eventually took as well as the role played by the concepts of EM and SD as well as the works by Giddens and Beck.

CATTON AND DUNLAP AND THE PROPOSAL FOR AN ENVIRONMENTAL SOCIOLOGY

At the end of the 1970s, Catton and Dunlap published two articles presenting an ecological criticism of contemporary sociology and outlining a proposal to create an environmental sociology. In those two articles are found the main arguments and some of the most polemic ideas from those authors on the relationship between sociology and the environmental issue. Due to this, we will mainly refer to the arguments put forward in these two articles where the idea of an environmental sociology was launched by Catton and Dunlap.[1]

The starting point for Catton and Dunlap was the criticism they addressed both to classical and the contemporary sociology. Basically, those authors pointed out the absence of any concern with the ecological base of the society

7

throughout the history of sociology. At the same time, they argued that the tendency to neglect physical and biological factors of the environment was even seen as a sign of maturity in the development of the social sciences. At the end of the 1970s, Catton and Dunlap had doubts about the certainty of such success. According to them, the progressive substitution of deterministic explanations connected to the physical environment through sociocultural explanations also led sociology to ignore environmental factors that are embedded in social life. According to their observations, human beings are connected in an interdependent way to other species in the web of nature.. In their 1978 article "Environmental Sociology: A New Paradigm," they argued that the various theoretical perspectives prevalent in sociology, when viewed through the ecological lens, are not very different. These theoretical perspectives include functionalism, symbolic interactionism, ethnomethodology, conflict theory, Marxism, and others. For Catton and Dunlap (1978, 42), there is an anthropocentrism that incorporates all these existing approaches in sociology that make them all insensitive to the environmental question in some way. Such a general anthropocentric vision was called the human exceptionalism paradigm (HEP). From this follows the need for a paradigm shift, for the incorporation of environmental issues into sociology would tend to change all these theoretical perspectives. Thus, challenging sociological anthropocentrism would lead to a revision of existing assumptions in schools of thought in sociology.

Catton and Dunlap blamed the set of anthropocentric suppositions contained in the HEP for the difficulty sociologists have found when approaching ecological issues. An evolutionary sociological theory such as Parsons's, according to Catton and Dunlap, seldom shows "attention to the resource base" (1978, 43). In in Parsons's theory, Catton and Dunlap noted, nature is subjected to more efficient exploitation as societies become more differentiated internally and are thereby "adaptively upgraded." And the idea of the environment, both in this and other sociological theories, is reduced to the idea of a symbolic or social environment. It would be difficult, then, for sociologists or those who remain faithful to the suppositions contained in the HEP to consider the "natural laws" that have influenced social life and take into account the "carrying capacity" of the environment. And even when sociologists consider such aspects, they perform it in a way that establishes a supposed elasticity of the carrying capacity and the stock of natural resources, suppositions that today, according to them, must be seen as questionable.

As an alternative to the HEP, Catton and Dunlap proposed a new set of suppositions that would make sociology more sensitive to the environmental reality, which they called the new environmental paradigm (NEP). Its principles were extracted from several papers produced by a small number of environmental sociologists in the 1970s. In this work, they defined environmental sociology as "the study of interaction between the environment and society" (1978, 44). As they recognized, this definition could be found in the work of others of other sociologists such as Allan Schnaiberg. From this concept, the study of environmental sociology involves studying the effects of the environment on society and, in turn, also studying the effects of society on the environment. Further to such a very general definition, it was not presented as another attempt to delimitate the concept of an environmental sociology. This only occurred in the following year, when Catton and Dunlap reformulated and gave continuity to their ideas.

In an article from 1979, Catton and Dunlap tried to provide a classification of what sociology had already produced in relation to the environmental theme, thus inaugurating a distinction between "sociology of the environmental issues" and "environmental sociology": The first is a tributary of more traditional sociology, only marginally incorporating the environmental theme, while the second has the ecological issue at its core. Added to this, this article contains a reformulation of the HEP versus NEP binomial. The initials are replaced by new meanings: human exceptionalism paradigm (HEP), and new ecological paradigm (NEP). In this new distinction, Catton and Dunlap (1979) took an apparently more anthropocentric position. The new ecological paradigm of Catton and Dunlap was presented as an alternative to the existing sociological paradigm. The latter assumes, among other things, that (a) humans are unique among living things on earth, (b) culture can change indefinitely, (c) human differences are always socially produced, and (d) cultural change is seen as continuous progress. Given these postulates from HEP, Catton and Dunlap presented three alternative postulates for environmental sociology; among these were the theses that (a) humans are not unique but one species among many others that are part of the web of life on the planet, that (b) the interdependencies that exist between society and nature tend to create unintended consequences for the intended actions of humans, and finally that (c) given that the world is finite, there are limits to human progress, particularly for material and economic progress. HEP and NEP were presented as opposing paradigms because these postulates could be relevant to sociological research: NEP postulates make sociology aware of the environmental issues, while HEP postulates make it indifferent to the subject.[2]

Catton and Dunlap recognized that sociological studies on the environmental theme have appeared in works based on traditional approaches and themes of sociology. They wrote that many sociologists would have been "drawn into the study of environmental issues through an interest in traditional sociological areas such as leisure behavior, applied sociology, and social movements" (1979, 246). For Catton and Dunlap, environmental sociology remained a quite broad conception, of which "the study of interaction between the environment and society is the core" (251). At the same time, the authors recognized that such interactions are very complex and varied, and this means sociologists must investigate a quite diversified range of phenomena. Thus, they propose an analytical structure inspired by the concept of the ecological complex, structurally interrelating the population, organization, and technology. Two concepts that have become of vital importance in that context are the ecosystem and the carrying capacity of the environment. This analytical structure would provide the means to assess the different existing levels in the relationship between society and the environment.

What Catton and Dunlap came to denominate the "ecological complex" found its inspiration directly in biological thinking and ecology. As they asserted, this analytical structure was developed "from the biologists' concept of ecosystem by Duncan . . . as part of his effort to apply insights from general ecology to sociological human ecology" (1979, 251).

A DEEP ENVIRONMENTAL SOCIOLOGY?

Catton and Dunlap suggested in their works that the studies that come from traditional sociological areas would serve as strong influence on the appearance of the environmental sociology. They argued in this article that it was the "sociology of environmental issues" that would lead sociologists to consider the ecological significance of social facts, setting the stage for them to move toward a truly ecological kind of environmental sociology. They even cited some areas of research in sociology (such as on natural resource management or wilderness re-creation) as those that brought about this change. Research in these areas, they claimed, has finally produced a "truly environmental sociology" (1979, 248).

This demonstrates that what Catton and Dunlap defined as a disruption of environmental sociology with traditional sociological approaches represents a trend, already existing in the "sociology of environmental issues," that privileges certain types of environment to the detriment of others (such as natural resources and wild areas). This trend does not represent, then, a rupture in the way nature is represented, but rather, reiterates an already existing vision that has been kept at the margin of sociological studies.. If the proposal by Catton and Dunlap were strictly followed, we should then recognize that environmental sociology is in fact not a study with major interest in the relationship between society and the environment but is interested instead in a more specific relationship: that occurring between society and natural environments. Environmental sociology, then, began expressing a trend toward seeing studies on the scarcity of natural resources, natural disasters, and wild areas as more important than studies, for instance, about the built environment. While the former deal with a more natural environment, the latter are imbued with the idea of a more artificial nature. In this case, the studies on natural resources have not only given the emergence to environmental sociology but also came to be seen as the most representative studies in the area.

This brings us to a fundamental aspect of the work of Catton and Dunlap. What calls the most attention to the distinction they established between HEP and NEP is the explicit bond between their ideas and some contemporary aspects of environmental thinking. The authors incorporated some ideas from environmental thought and tried to adapt them to a sociological perspective. For example, Catton and Dunlap's (1978) idea that human beings live in a finite world and that there are, therefore, physical and biological limits to human practices cannot be considered either original or modern.

The debate on limits in terms of resources or of the carrying capacity of environmental systems, in more general terms, has a long history dating back to the seventeenth and eighteenth centuries. For example, in 1650 the British doctor William Petty (1623–1687) had already speculated on the possibility that as a result of population growth, within a period of two thousand years, humanity would exceed the capacity of earth's ability to sustain it (McCormick 1992). However, it was Thomas Malthus (1766–1834) who became directly associated with this theme. In his 1798 "Essay on Population," Malthus argued that there was a discrepancy between exponential population growth and food production. In 1968, this issue was revisited by Garret Hardin in his article "The Tragedy of the Commons," which alluded to the possible compromising of the environment's capacity to support human life due to unlimited population growth. With this article, Hardin also sought to draw general attention to the existing antagonism between population and environmental resources.[3]

In 1968 and 1972, two other works were published with very similar argumentative structure to that used by Malthus and Hardin. These works were *The Limits to Growth* and *The Population Bomb*. The first made allusions to the possibility of a global environmental collapse due to the incongruity that, according to the authors, was being created between growth (population and economic) and the planet's resource base. *The Population Bomb* primarily focused on population growth and its impact on the capacity limits of the planet. As we can see, the idea that human practices are likely to threaten the environment's "carrying capacity" is far from being reduced to only the works of Catton and Dunlap.

Works such as Ehrlich's *The Population Bomb* and Meadows and colleagues' *The Limits to Growth* (1972) were not much more than a new version of old Malthusian ideas. In the words of Paehlke: "*Limits* was essentially a computerized Malthusian perspective applied to an industrial society, rather than to an agricultural society. Even more indebted to Malthus was Paul Ehrlich's sensational *Population Bomb*, published in 1968" (1989, 42).

However, the assumptions that Catton and Dunlap (1979) established for environmental sociology are attributed not only to authors such as Malthus and works such as *The Limits to Growth*, but also to a more romantic facet of contemporary environmental thinking. This romantic matrix is more clearly revealed if we examine the link between Catton and Dunlap's ideas and deep ecology.

The foundations of deep ecology were established by the Norwegian philosopher and thinker Arne Naess, who created the name and set the main guidelines for this field of environmental thinking. Deep ecology is a version of contemporary ecological romanticism whose main interest is to develop new forms of subjectivity in order to provide an alternative means for individuals to experience nature.[4]

The link between deep ecology and the work of Catton and Dunlap can be seen in one of the assumptions they used to form their new ecological paradigm. The assumption asserts that "human beings are but one species among the many that are interdependently involved in the biotic communities that shape our social life" (Catton and Dunlap 1978, 45). This principle, which is part of Catton and Dunlap's new ecological paradigm, differs very little from one of the central pillars of deep ecology: biocentric equality. The latter assumes that "no species, including the human species, is regarded as more valuable or in any sense higher than any other species. The effective opposite of biocentric equality is anthropocentric arrogance" (Dryzek 1997, 156).

In Dryzek's work *The Politics of the Earth*, we can find a more general assessment by the author about the currently existing environmental discourses. "Discourse" for Dryzek (1997, 10) means the shared mode of learning the world, always steeped in some kind of language that enables those who subscribe to interpret the reality from fragmented information from the social and natural world, generating stories and consistent assessments of reality. All discourse provides assumptions, judgments that support agreements and disagreements on that same reality.

Survivalism is an environmental discourse guided by the idea that human demand on the ability of ecosystems to support life are threatening to go out of control, consequently demanding severe actions in order to curb this trend (Dryzek 1997, 34). The assertions of Catton and Dunlap that the "world is finite" and that, therefore, there are physical and biological limits that go against human practices are part of the environmental survivalist discourse, which can also be found in works such as those by Malthus and Hardin and in books such as *The Population Bomb* and *The Limits to Growth*. Dryzek (1997) mentioned a book by Catton (*Overshoot*) in his work as an example of the survivalist discourse. That bond that exists between the work of Catton and Dunlap and *Limits to Growth* is also recognized by Buttel, for whom "Catton's dire predictions are compatible with a good deal of nonsociological work, such as *The Limits to Growth*" (1987, 469). It is also worth noting that Dryzek (1997) ended up classifying Catton not as a sociologist but, as a population biologist. On the other hand, the argument that advocates the establishment of equality between human beings and other species (biocentric equality) is one of the key ideas of deep ecology that Dryzek calls the green romanticism discourse.

There seems to be no doubt that Catton and Dunlap shared many of the ideas and assumptions that belong to different discourses of environmental thinking, especially those Dryzek (1997) calls survivalism and green romanticism. The relationship Catton and Dunlap established with areas they considered as representative of the "natural environment" and the importance attributed to these areas in environmental sociology very much resemble those concerns of deep ecology. After all, as one author observes with regard to the latter, "the general goals of deep ecology can be stated as the preservation of nature 'wild and free' and the limiting of the human impact on nature as the way to achieve this" (Barry 1999b, 14). A "genuine" environmental sociology, for Catton and Dunlap, should also examine "natural resources" and "wild areas." Thus, while the concerns of deep ecology resonate on a political level and in the individual and social experience, Catton and Dunlap incorporated them into the sociological plan.

Another factor that has strengthened this view of environment as "wild and free" was that Catton and Dunlap made use of biological knowledge and concepts such as ecosystem. To generate a unique structure able to assess natural and social systems from their concept of "ecological complex," the authors disregarded differences that should be considered between one and the other. i. At the expense of natural systems, as the anthropologist Bennet (1996) wrote, social systems are a projection of human behavior. This implies that the stability or change of a social system depends on the interests, desires, and practices of those who remain.

The idea of "natural environment" is counterproductive to the creation of a socioecological theory, although it seems quite attractive because it apparently allows the question of values and interests present in interpretations that make up the environment to transcend. Such a notion associates the environment with something that is beyond human culture, something that is not influenced by our choices and cultural practices. This illuminates it, allowing us see it as something totally free of our subjectivity and evaluative choices.

Social theory should not dismiss the reality of natural processes that do not depend on our decisions to be realized but rather try to see in its nature that focusing on targeting social theory and political practice would not be the most sensible thing to do. What social scientists should do is the reverse: take as a premise that there are no "value-neutral" readings of the environment as nonhuman nature (Barry 1999a, 20). It is important to understand that Catton and Dunlap's concept of untouched nature is a result of the influence of survivalism, deep ecology, and biological thinking itself. However, it tends to disappear in sociological evaluations of contemporary environmental issues. This will become evident with the approaches that will be examined throughout the work.

Catton and Dunlap (1978) have named sociological anthropocentrism as the main culprit responsible for sociological neglect of the environmental issue. However, the reasons for this neglect are multiple and diverse. Firstly, whereas the success of the social sciences was seen by many social scientists to be a direct result of the marginalization of explanations based on environmental determinism, for many social scientists, speaking about greening sociology risked falling into some new kind of naturalistic reductionism. These social scientists might think that by arguing about the influence of "the environment" and "environmental factors" on society, we would be running the risk of ignoring the importance of the "social" in the explanation of human behavior and its institutions. Within this framework is what authors like Benton and Redclift (1994) call the danger of the Trojan horse of biological determinism. This danger arises, they say, with the attempt to introduce environmental thinking (e.g., biology) directly into the conceptual framework of social theory. That is why we will find that socialists and feminists reject the ideas of "nature" and "natural," for fear that these concepts will be used as tools for justifying class and gender inequalities. This rejection is because the "natural" has always been used as a ruse to legitimize the social world as it presents itself to us at any given time.

Secondly, we have to recognize that the environment is generally seen as a topic and specific object of the natural sciences, and it connects to the division of labor that arose between the natural and social sciences during the twentieth century. As indicated by Martell (1994), the debate on the environment often involves discussions on the pollution of water and air and the extermination of animal species that apparently are not very familiar to sociologists. In addition, some aspects of the environmental debate are deeply troubling for many social science researchers, especially the fact that green discourse often takes a catastrophic and alarmist form, bringing regulatory implications that sociologists sometimes don't appear likely to incorporate.

Thirdly, it should be noted that the influence of Marxism was, for a long period, a very influential contemporary theory in academia in many parts of the globe. There seem to be certain problematic factors in attempting to approach environmentally themed Marxism. This is because many Marxists tend to view the environment as something typical of the bourgeoisie (post-materialist) interests of the middle class and therefore far from the more noble and pressing (materialist) interests of the working class. The mere fact that many environmentalists accuse industrialism, and not necessarily capitalism, of being one of the main culprits responsible for environmental impacts is reason enough for Marxist authors to view the cause suspiciously.

The optimism for progress that Catton and Dunlap (1978) attributed to classical and contemporary sociology seems to also have made a strong contribution. This optimism is present in almost all the classics of sociology. The basic ecological question for classical social theory was the source of environmental degradation. First, what caught the attention of the classics was the question of how premodern societies remained subdued by their natural environment and how modern societies managed to transcend these limits, distancing themselves from their "natural" origins. From that point of view, if the classics addressed environmental issues, they did it more so in order to investigate the reasons why modern societies would be getting rid of ecological pressures than to reaffirm our dependence on them. Another reason why the classics neglected the debate on environmental changes lies in the fact that industrialization (which was the subject of central concern to Marx, Weber, and Durkheim) contributed to the relegation of the dark forecasts of authors such as Malthus to the background. This seems to have been central to the position that the classics of sociology adopted in relation to the environmental theme.[5]

Many contemporary social scientists agree that sociology started too late in worrying about environmental issues.[6] But if on the one hand there is agreement between the work of Catton and Dunlap and contemporary social scientists, such similarities end there. On other points, it is possible to perceive the gap that occurs between what was produced by Catton and Dunlap and what has been more recently produced by other authors. As we have seen, Catton and Dunlap's work seems to have presented certain errors and ambiguities that the latest sociological works would hardly endorse. Firstly, however much that Catton and Dunlap committed themselves to creating a "new ecological paradigm," it seems they never made it very clear how this paradigm would be, considering the very tension created between sociological and biological thinking.

In addition, although it is part of the debate, the major problem with creating an environmental sociology does not reside in being centered on "anti-ecologic" anthropocentrism, which, to Catton and Dunlap, was guiding sociology in the direction of insensitivity to the issue, but in the existing division of labor between the social and natural sciences.[7] The fact that social scientists consider the environment as a specific theme of natural scientists and the attempt of the classics to separate sociology from biology were both the result of the division of labor that occurred between natural sciences and social sciences at the beginning of the twentieth century and also a result of a division of labor that was happening within the social sciences (the distinction between sociology, psychology, and economics). The conceptual matrix that sociology came to define for itself, a process that was, in turn, shaped by the existing competition between itself and disciplines such as biology and psychology, stimulated the exclusion of considerations in the substrate "material" of society in sociological thought (Benton 1994).

We have made these observations to arrive at two basic conclusions. First, while some authors agree on more general aspects of the work of Catton and Dunlap, others raise a number of problems. Breaking with sociological reductionism cannot be resolved at the expense of biological reductionism or carelessly incorporating the thought of the natural sciences with the social sciences, a problem that Catton and Dunlap did not satisfactorily elucidate. Secondly, although we may find similarities between the principles that Catton and Dunlap outlined for environmental sociology and the new ecological directions taken by contemporary sociology, there are fundamental differences in these new directions that must be worked on. Catton and Dunlap's attempt to green sociology was made by attempting to directly incorporate specific ideas associated with environmental thinking as well as concepts of biological sciences into the conceptual framework of sociology. But if the greening of sociology means a move of this type, as Catton and Dunlap sought to do, then we should ask ourselves why sociobiology could not provide the basis of ecological sociology.

Sociobiology is just a set of theories that, having arisen in contemporary biology, looks for biological theories to apply to understanding the human world.[8] The ideas contained in sociobiology already seem to seduce some social scientists, because they appear to be perfectly in line with the intention of seeking a more biological basis to the social sciences. However, appealing to the natural sciences as such implies open space to re-create a biological reductionism in the social sciences, much like that of the late nineteenth and early twentieth centuries: a reductionism that classical authors like Marx, Weber, and Durkheim were interested in breaking and that shaped a sociological critique of environmental determinism that can be seen as valid even until today.

What some authors call realism is often used as a way to resolve this impasse and to establish a more ecological social theory. In *Society and Nature*, Dickens (1992), for example, considers many of the important conclusions of Catton and Dunlap on the relationship between sociology and the environmental issue, including the authors' emphasis about the "materiality" and the "ecological constraints" in social life. But Dickens, along with Benton (1991), see the need to anchor these issues within the sociological imaginary of Marx and Engels as well as contemporary realist theorists such as Roy Bhaskar. Critical realism can be seen, then, as a result of a Marxist heritage that is developed more specifically by authors such as Dickens and Benton.[9]

Social theorists who advocate for environmental sociology realism argue that human beings and social groups keep a relationship, both materially and symbolically, with the environment. These dimensions are interwoven together. Benton (1994) and Barry (1999a) use the expression of human "embodiedness" and "embededness" to allude to this double process. These terms (embodiedness and embededness) suggest that the relationship between humans and their environment is both material and symbolic at the same time and that these dimensions are therefore somehow intertwined. A social realist theory does not disregard the unique capabilities of humans, nor does it diminish their collective capacity to express this specialness. On the contrary, a social theory insensitive to the unique capabilities of humans that does not recognize their condition of *being part* and, at the same time, *being a part* of the environment, as Barry (1999a) argues, would be unable to understand and recognize the variety and complexity involved in the relations of the different human groups with their environment. Realism is important because it allows us to stratify different levels of knowledge and to combine various disciplinary perspectives without necessarily reducing a type of knowledge (biological) to another (sociological). Realism, according to Dickens (1996), offers the prospect of integrating knowledge without us having to fall into a deep and incurable eclecticism. But, as the author himself acknowledges, realism reveals that mechanisms exist in the different strata (physical, biological, and social) and even tells us, right away, how they connect to each other.

ENVIRONMENTAL SOCIOLOGY: TASKS AND RESEARCH THEMES

Several authors have written about the tasks and objectives of environmental sociology research. When considering their arguments, we can outline three main areas of interest for environmental sociology: (a) social practices and environmental change, (b) knowledge and interpretations of the environment, and (c) ecological policy. Next, we will evaluate these proposals and relate them to the concepts of SD and EM and to the theory of SR.

Social Practices and Environmental Change

There is a consensus among social scientists that one of the main focuses of interest of environmental sociology should be the relationship between social practices and environmental change. One of the phenomena that should be explained by environmental sociology is precisely the impact that intentional and unintentional social practices end up having on the environment. For

Buttel (1996), among these daily practices should be included our practices in the process of production and consumption.

Other authors arrive at the same conclusion. According to Dickens (1996), the bigger problem is in the division of contemporary work and the intellectual division that it implies. This is because such divisions have fragmented the set of transformation practices of nature and also the knowledge of these activities. Modern societies transform nature within a framework of high labor specialization and on a scale that is now global. In this way, industrial and consumer practices are key issues for a sociology concerned with environmental issues. Macnaghten and Urry (1995, 1998) and Hannigan (1995) speak to us of "environmental destruction" and of environmental "danger." According to these authors, the analysis of social practices can significantly contribute to the understanding of processes that currently produce what are recognized as environmental dangers.

Knowledge and Interpretations of the Environment

There are some basic aspects raised by sociological literature concerning the issue of environmental knowledge. The first of these concerns the situation in which an environmental change is recognized as an environmental "danger," or "risk." As we saw earlier, one of the objectives for environmental sociology is to investigate the way in which social practices end up creating environmental impacts. However, if the analysis is left at this level, it does not respond to some very important issues.

Firstly, why should certain environmental changes be seen as "dangerous" while others are not? When should environmental changes be seen as entailing unacceptable risks? Do all societies or social groups react similarly in relation to human intervention in the environment? Some authors seek to answer these questions as follows: The awareness that we have of environmental problems is a direct result of the impact that we created to the environment.[10] In this way, the environmental movement would be a direct result of pollution. Hannigan (1995) calls this the thesis of reflection. But as he and others point out, this vision could prove fragile, as the concern with environmental problems can exist independently of the magnitude of its problems, as there are values and a cultural context influencing the perception of our intervention in the environment and our reaction to this intervention. In addition, the idea that our concern with environmental changes is a direct reflection of our intervention in the environment can be put in check by the fact that many of the impacts we have created (e.g., genetic mutation, acid rain, climate change, etc.) are virtually invisible to the perceptual organs of an ordinary person. If environmental awareness were mere reaction, why bother people with issues such as these if we cannot understand such phenomena in our daily life?

This shows not only that scientific knowledge is a central variable for public recognition of the existence of certain environmental problems but also that their dissemination through mass media has a fundamental role. As public recognition of these problems is mediated by some sort of knowledge, one of the objectives of environmental sociology has become to evaluate the various ways in which the environment can be seen by social groups and the different ways in which an environmental problem can be defined. Environmental sociologists should, then, be concerned with the different "cultural readings of nature" (Macnahten and Urry 1998) and search for the various ways by which the environment is perceived and assessed by social groups, be they entire societies, communities, social movements, research institutes, and the like.

All this is synthesized by the argument put forward by Buttel and Taylor (1994) that environmental sociology must encompass a sociology of knowledge. The recognition that interests and values interfere in scientific assessments is constantly approached by this research area of Sociology.[11] This perspective is also consistent with the proposal of Barry (1999a), related in the previous pages, that a social theory should not admit the existence of a *reading-off* of the environment. This stance is also present in the work of Giddens (1991) and Beck (1992), who warn that any ecological risk assessment involves some kind of value judgment.

Environmental Politics

As Yearley (1992b) argues, modern environmental threats are of two distinct forms. At first they appear as environmental changes that may or may not bring serious consequences to human beings. Threats can arise also in an "ideological" form, usually at the hands of the environmental movement. So while environmental problems involve some kind of "physical" change, the environmental challenge as a set of values and ideas is the ideological content in search of a broad institutional change to society.

In a sense, environmentalism as a social movement arises as a reaction to the growing human intrusion in the environment, and it is seen as a reaction to the human destruction of the environment that makes it necessary, at the very least, that we take certain precautions with deterministic views of the clash between society and the environment. Theories that assume an inevitable trend of modern societies toward ecological crisis as their starting point can fall into a kind of determinism that makes it impossible to assess changes that comprise an environmental improvement. Buttel (1996) calls us to pay attention to those theoretical systems that emphasize the immutability of the forces leading to environmental degradation, introducing a characteristically deterministic profile of these changes, thereby being unable to explain the conditions under which positive socioenvironmental changes are likely to occur.

Hannigan (1995) notes that part of the environmental sociology literature, specifically that related to the contemporary ecological Marxism, has produced a monolithic vision of the state. The latter is generally seen as one of the leading promoters of environmental destruction. Social scientists, therefore, tend to overlook the role of the state in carrying out ecological policy and end up putting much of their hope in the environmental movement. There is a misunderstanding here, as this posture disregards one of the complexities that arise with state intervention in environmental care: their dependence upon and the routine use of scientific knowledge in attempting to define environmental protection. However, this dependence on scientific knowledge is something that is also characteristic of the environmental movement, which means that not only the state but also the environmental movement itself risks being overrun by possible contradictions, when it tries to sustain its arguments on a "scientific basis."[12]

In turn, Dickens (1996) provides us with an alternative reading. For him, one of the basic reasons for our lack of understanding of environmental problems lies in the division of labor. It is precisely the latter that fragments our understanding of nature. Faced with this division of labor, the state has an important role: As the division of labor expands, it demands some form of controlling or coordinating role that could be played by the state, which, in this case, would function as an organizing agency of environmental knowledge.

NEW DIRECTIONS OF ENVIRONMENTAL STUDIES

In the late 1990s, Catton and Dunlap recognized not only that their proposal for environmental sociology had failed, but also that the situation of environmental sociology had not changed significantly in relation to the framework they found in 1970s. However, other authors did not seem to agree with that assessment, claiming that there was not a decrease in sociological studies on environmental issues, but, on the contrary, that there was a stunning increase of these studies taking place both within and outside of the social sciences. This seemed to indicate the emergence of an ecological cacophony. Some of these authors also note that instead of stagnating, today there is the possibility of environmental sociology studies expanding its horizons beyond the research objectives Catton and Dunlap formulated in the 1970s and 1980s. Some social scientists mention the possibility of re-creating environmental sociology, while others suggest that it seek new directions. The concepts of

EM and SD along with the works of Giddens and Beck are usually associated with this redirection of environmental sociology.[13]

So far, there are few efforts that provide an overview of environmental research currently developing in the social sciences. In addition, some of the existing classifications are more general than others and have at times shown differences with respect to the inclusion of certain approaches in the social science research field.[14] Bryant and Bailey (1997) note that the environmental theme is now being incorporated into a variety of disciplines within the social sciences. Thus, new disciplines are emerging that give intellectual expression to these new interests. These include cultural ecology, ecological economics, environmental history, environmental management, environmental policy, environmental sociology, global ecology, and others.

In addition to these new disciplines, new areas are emerging within certain fields of knowledge, leading to internal knowledge conflicts. An example of this is economics. In this field there is, on the one hand, what some authors call "environmental economy" and, on the other hand, "ecological economy." Some authors consider the latter to be more "radical" than the former. In the field of environmental sociology it is no different.[15] It is also possible to find subdivisions within some of these approaches. In the field of sociology of risk, Rosa (2000) presents at least four different strands.[16]

Strydom (2002), in turn, distinguishes sociological approaches that address environmental issues on the basis of the realism/constructivism axis. For him, sociological approaches are arranged along a line from strong realism to strong constructivism. The distinction Strydom makes between these different theories seems questionable in some respects. The structuration theory of Giddens (1984) is considered by others to parallel the critical realism of Bhaskar.[17] In turn, EM theorists such as Mol (1995) attempt to approximate Giddens's structuration theory. Accordingly, it might be a mistake to assign Giddens to "weak constructivism" and EM theorists to "strong realism" because they are so close. The problem is that some of these perspectives, such as Giddens's theory of structuration, attempt to overcome a narrow division between realism and constructivism. This is perhaps why Strydom attributes to Giddens both "weak constructivism" and "constructivist realism."[18]

All this seems to show that environmental sociology rests in irremediable eclecticism. It seems to have become a less consensual area of knowledge than it was one or two decades ago. Today we can find a plethora of theoretical trends invading the social sciences with relation to environmental issues. We will not focus in this work on the positive and negative aspects of this picture. Perhaps at this time it's important to remember that such diversity has been a characteristic aspect of sociology since its emergence (Stehr 1982). This issue will be partially treated when we discuss the concept of SD, which is justly criticized for the conceptual diversity that it expresses. Anyway, this group always puts forth some important questions for the social scientist and researcher: What are the most promising approaches to the understanding of modern environmental problems? Are these approaches so different from each other to the point of making a synthesis between them impossible?

The analysis of the similarities and compatibilities between the different theoretical chains of environmental sociology deserves greater attention from social scientists. In the field of sociology of risk, Rosa (2000) mentions the possibility of a reconciliation of different theoretical perspectives. Buttel (1996), in reference to the growing theoretical diversity of environmental sociology, argues that the possibility of synthesis seems to be unfeasible due to differences that are highlighted by observers. However, this debate on the synthesis neglects, as we are reminded by Buttel, the analysis of specific issues in giving greater attention to the superiority or inferiority of one or another theoretical system.

It is critical to assess to what extent the existing perspectives on environmental sociology are different from each other and whether this difference makes them incompatible. Likewise, it is important to evaluate to what extent these perspectives feature points in common or how their possible differences can contribute to rather than prevent a more complex and general understanding of the socioenvironmental reality.

It is not our goal in this work to investigate all problems arising from the meeting of these different theoretical chains of environmental sociology. Our first aim is to outline our analysis in three perspectives that are considered vital to the greening of sociology in recent years, showing how different theoretical perspectives can bring individual contributions to environmental sociology on crucial issues and problems for the area. These prospects are EM, SD, and the theory of RS. These approaches have already been presented preliminarily in the introduction, but it is appropriate to return to them not only to put them in perspective to the internal debates of environmental sociology that we have just examined, but also to access, in more detail, the problems and questions that arise of the encounter that takes place between them in literature. In the remaining part of the book, we will examine some of these issues in detail.

ECOLOGICAL MODERNIZATION, SUSTAINABLE DEVELOPMENT, AND RISK SOCIETY

The prospect of EM is in line with many of the interests that are outlined for the agenda of environmental sociology. Firstly, it is compatible with realism. Thus, EM theorists take the evaluation of "flow of substances, flow of energies, the movement of materials through human societies, etc." as a central task (Mol, Spaargaren and Buttel 2000, 6). In addition, EM theorists have conceptualized the relationship between society and environment in order to avoid a type of biologism about which authors such as Benton and Redclift (1994) warn us. Although EM theorists establish interdependencies between society and the environment, they accept the existence of different rationalities (ecological, social, and economic) governing this relationship. Social and ecological systems are thus not fully diluted by each other, although linkages between them can be established.

But it must be pointed out that EM takes different forms. According to Mol (1995), three different uses of the term can be found. First, it can be seen as a new concept that brings theoretical contributions to a new branch of sociology—environmental sociology. A second strand sees EM as set of social science studies that seek to analyze environmental policies that encourage a greener pattern of production. In this case, EM is understood as a new ecological discourse that would lead to a new paradigm of environmental policy. A third chain considers EM as a concrete program of environmental policy set in motion by political parties.

In its sociological dimension, EM theory provides a set of concepts aimed at understanding the origins of modern environmental degradation and evaluating how organizations respond to these problems. Mol (1995) sees EM as an industrial transformation process that enables the promotion of environmental sustainability (or the "support base," in his words). Mol adds that from this understanding, EM proposes the possibility of overcoming the environmental crisis through an institutional transformation in which the institutions of modernity are not eliminated but reconfigured without having to abandon the pattern of Western modernization. In short, EM means preserving the ecological basis of the industrial system by creating a process of institutional innovation that sustains it, and this process can be seen as triggering a new wave of modernization.

The importance of EM for environmental sociology seems to reside in both the importance it gives the possibility of integrating economy and ecology and also the importance it attaches to the state as the "driver" of this change. As we will see later, the changes declared by EM establish a strong presence of the state, both to trigger the integration of economy and ecology and to fill the gaps and shortcomings in carrying forward this process. But it must be pointed out that while it establishes a role and a type of action for the state in generating environmental policy, EM emerges from the thesis of state failure with regard to its performance in environmental regulation. In a way, EM is part of the critique of the state's fragmented, bureaucratic, and reactive ecological policy from the 1970s.

Today, SD and sustainability are terms known worldwide. Although SD is a relatively new term in the vocabulary of politics and contemporary social sciences, its origins date back to the beginning of the last century. As we mentioned in our discussion on the environmental sociology of Catton and Dunlap, concern about the limits that industrial growth and population put on the environment is not new. However, this concept only explicitly entered into the scenario of global concerns with the publication of the report *Our Common Future* (WCED 1987). Since then, SD has become a more and more widespread term within the social sciences and in environmental conferences involving rich and poor countries worldwide.

The concept of SD combines an interest in the environment and environmental protection with obligations to present and future human generations. In the view of authors like Barry, there are several aspects of the discourse surrounding SD that are in line with socioecological theory. Among them is concern for (a) human dependence in relation to the natural environment, (b) the existence of external natural limits on economic activity, (c) the pernicious effects of certain industrial activities on local and global environments, (d) the fragility of these local and global environments in relation to collective human action (e) the recognition that initiatives related to development should be linked to their own environmental preconditions, and (f) decisions on development and its consequences for future generations and for those who live in other parts of the planet (Barry 1999a). Thus we see that SD, much like the EM discourse, seeks to promote integration of economic interests with environmental demands. On this point, the concept of SD is very similar to that of ME; both concepts share a vision that tends to see as possible the reconciliation of economic activities with the capacities of ecological systems.

With respect to the RS, Giddens and Beck are considered the sociologists who have contributed most to approaching the ecological subject in environmental sociology.[19] Their works are considered an important starting point for understanding modern environmental degradation and the changes and conflicts associated with it. For example, for Goldblatt (1996), the work of Giddens and Beck would make it possible to place the origin and consequences of environmental degredation at the center of the concerns of social theory.

It is worth highlighting three basic points in relation to the work of Giddens and Beck. The first of these refers to the issue of environmental risk. Giddens and Beck seek to highlight the global aspect of the threats we create to the environment and to humans. This question is assessed in the work of the authors from the discussion on the emergence of high-consequence risks and of the change in the "risk environment" from the premodern to the modern context. Secondly, along the same line of the debate about the tasks and objectives of environmental sociology research, both Beck and Giddens underscore our dependence on scientific knowledge with regard to

environmental problems. Thirdly, both seek to extract the political consequences of these changes in contemporary societies.

Beck and Giddens clearly converge in their considerations on the emergence of risks with high potential impact and on their implications for the
emergence of what they term reflexive modernity. For Beck (1992), the first
phase of modernity is represented by the emergence of industrial society,
which had the question of production and distribution of goods as its organizing principle. Beck and Giddens (1991) both point to the emergence of a
second phase in modernity, marked by the emergence of RS: a society that
has risks instead of the distribution of goods as its axial axis. Among these,
the ecological risks are the most emblematic in that change for both Beck
and Giddens.

In the next chapters, a more systematic analysis will be conducted of each
of these approaches in order to unravel the controversies that arise in the literature about the similarities and differences between them. As we have seen,
SD, EM, and the RS theory bring forth some contribution to environmental
sociology. However, these prospects are seen, at certain times, as conflicting
or even diametrically opposed to each other. Thus, there are disagreements
between social scientists about the contribution these concepts may or may
not bring to environmental sociology. The concept of SD, for example, is
shrouded in controversy; one of them is linked to the possibility of reconciling development (or economic growth) with environment. In addition, an
aspect that draws attention relative to the concept of SD is the diversity of
visions and interpretations that surround it. This problem is absent in the concept of EM, according to some of its supporters. For many social scientists,
the interpretative diversity that imbues the concept of SD is shown as a drain-
ing aspect of the concept. That makes it a cliché and justifies the contention
surrounding the idea of SD; this, in turn, precludes putting a coherent environmental policy in motion. On the other hand, EM and SD are sometimes
considered quite similar in their approaches to integration between economy
and ecology, for which some authors see the first as a replacement of the
second. EM could be seen, in this way, as a more conceptual variation of SD,
increasing interpretative diversity that involves the concept.[20]

In contrast, other authors feel the need to define differences between SD
and EM. EM, according to these authors, is an overly narrow perspective
on two basic points: firstly, in its geographical scope—EM does not address
international issues and dilemmas posed by global environmental problems,
which are precisely the major concerns of the contemporary environmental
debate—and secondly, that EM is overly restricted in its political and moral
content. Thus, it is accused of being too technocentric and economic. It does
not take into account the relationship of environmental crisis with issues
involving future generations, rich and poor countries, or our relationship with

other animals. Some of these issues, it should be noted, are considered as strengths of SD's discourse.[21]

Differences are also noted among either EM on one side and RS theory on the other. Some authors situate the EM theory in an opposite perspective to the RS theory of Beck: They provide different ways to interpret intrinsic processes toward society and the environment. EM defends the possibility of accommodating environmental issues within the process of capitalist production and consumption, while the RS theory, taking a more critical attitude to modernization, tends to analyze the environmental crisis with greater depth.[22]

With regard to differences between these perspectives, it is worth noting that unlike EM, for the RS theory the environmental theme is addressed with a more global perspective than one at the national or regional level. The global character of modern pollution, which Giddens and Beck emphasize in their discussion of high-consequence risks, becomes important because it demystifies the attempt to reduce ecological policy strictly to national terms. Finally, and perhaps by virtue of the characteristics of high-consequence risks, these two authors make a more modest assessment regarding the role of science and technology in the context of environmental changes.

CONCLUSION

Throughout the present work we intend to assess to what extent these differences mark the dialogue between SD, EM, and RS, and put them in conflict with each other. In the subsequent chapters we will show that while these approaches present distinct profiles, environmental sociology has a lot to lose if we conform with rigid placements, since each of them brings a specific contribution to environmental sociology, focusing on topics and issues that are central to that specific area. At first, the differences of emphasis about the issues and the problems each present could be interpreted as exclusionary and unviable for competing theoretical perspectives; however, this impression can be undone if we better analyze the similarities and differences between these approaches. This is the axis of reflection that will unfold in the following pages.

NOTES

1. Their arguments are also presented in Catton and Dunlap (1980), but there are no significant changes to their arguments in this paper.

2. For the authors' arguments on these paradigmatic differences, see Catton and Dunlap (1978; 1979).

3. Contemporary authors such as David Ricardo, John S. Mill, and Karl Marx also engaged in this debate at certain times in their lives. On this point, see Tamanes (1985) and Benton (1991). An important aspect regarding the Tragedy of the Commons was the author's observation that the issue of the commons was not new and was already well known in social science circles (McCormick 1992). An example cited by McCormick of a work from the social sciences that addresses problems such as these is *The Logic of Collective Action* by Mancur Olson.

4. For an assessment of the emergence and development of deep ecology thinking, see Dobson (1990) and Barry (1999b).

5. Some of these arguments can be found in Barry (1999a), Martell (1994), and Goldblatt (1996). For an analysis of Marx's critique of Malthus, see Benton (1991). Although Buttel (2000b) agrees with Goldblatt about the optimism that prevails in classical sociology about these issues, for him classical sociology was much more ecological than the mainstream of contemporary sociology.

6. This diagnosis can be found in authors such as Goldblatt (1996), Dickens (1992; 1996), Giddens (1990), Beck (1992), Eder (1996b), and Machgnaren and Urry (1998).

7. One could say that this argument was already present in the work of Catton and Dunlap (1978) when the authors criticized Durkheim's point of view defining sociology as the science of "social facts." However, not only did Catton and Dunlap not develop this argument, but the answers they attempted to give to the problem were not satisfactory. For this argument, see Dickens (1992).

8. For an assessment of the impact of sociobiology on the social sciences, see the article by Nielsen (1994).

9. On the topic of realism and its role in the construction of an ecological social theory, see Dickens (1992; 1996), Benton (1991), Barry (1999a), New (1995), and Redclift and Woodgate (1994).

10. For an assessment of the different theses on the emergence of environmental awareness, see Hannigan (1995).

11. Regarding the importance of environmentalism for the sociology of knowledge and the contribution of this area to environmental issues, see the works of Yearley (1995) and Wynne (1994).

12. On the dependence of the environmental movement on scientific knowledge and the consequences this has for the movement itself, see Yearley (1995, 1992a).

13. Some of these arguments can be found in Buttel (1987, 1996); Cohen (2000a, 2000b); Gramling and Freundeburg (1996); Martell (1994); Benton and Redclift (1994); Barry (1999a); Lash, Szerszynski, and Wynne (1996); Mol, Spaargaren and Buttel (2000); and Gramling and Freundeburg (1996).

14. For some of these classifications, see Bryant and Bailey (1997), Pardo (1998), Strydom (2002), and Rosa (2000).

15. Pardo (1998) has included the following existing approaches in environmental sociology: (a) new ecological paradigm of Catton and Dunlap, (b) deep ecology and the Gaia hypothesis, (c) social ecology, (d) ecological modernization, (e) ecofeminism, (d) sociology of risk, and (e) society of waste.

16. The different research strands presented by Rosa (2000) are (a) Durkheimian tradition (Douglas and Wildavsky), (b) Marxist and Weberian tradition, (c)

utilitarianism and the rational actor paradigm (RAP), and (d) phenomenological tradition.

17. The works of Cohen (1989), Bryant and Jary (1991), and Kaspersen (2000) deepen this argument.

18. The debate between realism and constructivism has proven fruitless in debates within sociology itself, and it is unlikely that debates in environmental sociology will be any different. Thus, environmental sociology will be doomed to failure if it attempts to rely on a simplified version of each of these options. For an analysis of realism and constructivism in the social sciences, see Delanty (1997).

19. For assessments of the contribution by Giddens and Beck to the greening of social theory, see Hannigan (1995); Goldblatt (1996); Cohen (1997, 2000a); O'Brien, Penna, and Hay (1999); Lash and Wynne (1992); Lash, Szerszynski, and Wynne (1996); and Dickens (1992).

20. To elaborate on these arguments, see Redclift (1987), Sachs (1993), Lélé (1991), and Boland (1994).

21. See Christoff (2010) and Blowers (1997).

22. Some works that point to this contrast are Blowers (1997), Cohen (1997), and Mol (1995). We could also say that the conflicts and dilemmas that permeate the relationship between EM and RS can be extended between EM and RS can be extended to the relationship between EM and Anthony Giddens' sociology of modernity. Indeed, the latter considers it possible to describe modernity in terms of the theory of RS.

Chapter 2

Ecological Modernization

Economic Growth versus Environmental Protection

This chapter will address two distinct views of EM. One view considers EM an ecological discourse and another defines it as a sociological theory. However, the difference between these two approaches will serve only as a point of departure. Our main aim is to focus on a general trait common to both views, which permeates the discursive and sociological condition from EM. While the first part will be a brief evaluation of this dual condition of EM, the rest of the chapter will focus on an aspect central to both views. An assessment of the implications of the issues raised in this chapter for the formation of environmental sociology will end our reflection on EM. In the evaluation of EM as a discourse, we will focus on the works *The New Politics of Pollution*, by Albert Weale (1992), and *The Politics of Environmental Discourse*, by Maarten A. Hajer (1995). We will refer sometimes to the work *The Politics of the Earth*, by John Dryzek (1997). There is a consensus among these authors considering EM as a new type of environmental discourse (or ideology). In this line of interpretation, EM is a set of assumptions that points to a disruption in the recent development of European environmental policy. However, for the sense of EM as environmental sociology, we will use the work by Arthur P. J. Mol (1995) and Gert Spaargaren (1987). These authors are known as central figures attempting to elevate EM as the condition of environmental sociology.[1]

THE ORIGIN OF THE ECOLOGICAL
MODERNIZATION DISCOURSE

We cannot understand the discourse surrounding EM unless we refer to the factors that enabled its emergence in the 1980s. As a political discourse, EM did not emerge by chance but was the result of a series of changes taking place since the 1970s that created a favorable social context for its emergence in the following decade. Its emergence is thus linked to the reflections that took place in the 1980s with the aim of criticizing the decision-making processes of the environmental policies implemented in the 1970s. Their emergence would have been impossible, according to Weale, had it not been for "an earlier generation of policies, laws, regulations, and institutions" (1992, 2).[2]

What were these changes that led to the emergence of EM? Weale (1992) and Hajer (1995) offer similar answers to this question. Many of these changes happened because environmentalism was reached by a "mixture of fellings" in the 1970s (Hajer 1995, 87). This ambiguity had its origin in the coexistence of two major contrasting trends that prevailed in the environmental movement. Works such as *The Limits to Growth* (Meadows et al. 1972), which came to have a strong impact on the environmental movement of the period, began to emphasize the need for a greater input from science and technology. Works such as *A Blueprint for Survival* and *Small Is Beautiful* were also very influential among groups starting in the opposite direction. Unlike *The Limits to Growth*), these last two works began a general criticism of consumer society and specifically of the excessive confidence placed in technological innovations. According to Hajer, this led to a mixture of feelings between the environmental groups leading strategic change. Thus, in the early 1980s, the profile of the environmental movement was no longer the same. While in the 1970s the movement was characterized by a permanent attitude of confrontation with the state, in the 1980s it started to become "less radical, more practical, and were much more policy-oriented" (1995, 93).

Several factors caused this change in the posture of the movement. The economic recession that hit European countries in the late 1970s was one of them. As the economic theme reappeared with all its strength in the political agenda, environmental groups were forced to find a means of reconciling economic restructuring with environmental protection in order to regain public support for this discourse. From then on, it became important for the environmental movement to see the market economy and protection as partners and not as enemies. However, there are other factors involved in this strategic change in the environmental movement. Among them its growing professionalization. At as the environmental movement progressively became more professional, including in its staff specialists from diverse areas (engineering, biology, economics, and marketing), confrontational approaches used in the 1970s began to lose their meaning and to be themselves considered obstacles that made it difficult for groups to strengthen their political power (Hajer 1995, 94).[3] In addition, there has been a growing recognition of the failures of previously existing government environmental policies, further strengthening the EM discourse. The infeasibility of use these policies to address new transnational environmental problems that were emerging also became evident. As Weale (1992) indicates, it had become clear that the environmental policies of the 1970s had left several problems unsolved or even worse than before. This awareness was present not only in the bureaucratic elite of industrialized countries but also in the environmental movement itself.

Finally, it should be noted that the appearance of discourse of EM would not be possible, as Weale (1992) and Hajer (1995) argue, if an alternative environmental language had not emerged that would allow governments and other organizations to structure the problem in a new way. The environmental movement came to really change its political practice only because, as Hajer (1995) indicates, an alternative discourse was accessible. The language of EM started to emerge in various academic circles and in works that resulted from alliances between environmental NGOs and transnational organizations (the Organisation for Economic Co-operation and Development [OECD], the United Nations [UN], and the UN Environment Programme [UNEP]). It was because of this that, in the second half of the 1980s, as Hajer tells us, "ideas of ecological modernization had, by then, already overcome their growing pains. Work in academic circles and expert organizations now provided an alternative conceptual language and delivered concrete solutions could indeed be found" (95). Weale (1992) notes this same process. After all the setbacks that occurred with European environmental policy, according to him, a new belief system emerged that came to be called "ecological modernization." This belief system challenged the fundamental assumption of the political conventional wisdom that there was a zero-sum game between prosperity economic and environmental care. The activities carried out by the OECD and the UN appear to have been a major influence on the emergence of this new language. Many of the ideas that constitute the premises of ME's discourse were born from the activities promoted by these organizations. The proposal to see pollution as a question of inefficiency of industrial and technological systems, the argument of that the costs of pollution should be paid by the polluters themselves, and the belief in the compatibility between economic and environmental policies could already found in documents produced by the OECD. In the case of the UN, Hajer goes so far as to declare that the 1987 Brundtland Report (*Our Common Future*, produced by the UN), which popularized the concept of SD, "can be seen as one of the paradigmatic statements ecological modernization" (1995, 26). This view is shared by

Weale, for whom the central proposition of EM "emerged, most notably in the *Brundtland Report*" (1992, 31).[4]

All these factors contributed to some extent to the urgency of the EM discourse. However, most of them did not automatically generate that discourse. What they did was to provoke a repositioning of the existing political actors, forcing them to create and incorporate a new language of environmental policy. On the one hand, government groups have increasingly criticized the failures of the environmental policies implemented in the 1970s. On the other hand, changes that occurred in the 1980s came to further deepen the differences within the movement originated in the previous decade, which implied the adoption of a more pragmatic and cooperative stance by the environmental movement. Thus, as Weale makes clear, the "persistence and intensification of old pollution problems and the growth of new issues provided the occasion for a new politics of pollution to emerge in the 1980s" (1992, 28). This new pollution policy was, according Weale (1993), EM.

THE CENTRAL STORYLINE OF
ECOLOGICAL MODERNIZATION

Discourse has not been the only way to conceptualize EM. Weale (1992; 1993), for example, also uses the terms "belief system" and "ideology" to refer to it. For Weale (1993, 197), ideology (or belief system) is an interrelated set of concepts or propositions that have a dual function. On the one hand, it allows or provides a reference to describe what things in the social and environmental world are like, and, on the other, it prescribes how we should act in light of the descriptions that are made concerning these domains.[5] Hajer and Dryzek follow a similar line; however, they do not define EM as an ideology, but rather as an ecological *discourse*. For Dryzek (1997), discourse is a shared way of apprehending the world through stories or narratives that we create concerning the social and environmental world. Each discourse, like the conception of ideology by Weale (1992), is composed of assumptions, judgments, and statements that structure the different views that have emerged in the contemporary environmental conflict. The definition Hajer provides is very similar, and it is basically what we will refer to in subsequent parts of this chapter. He defines discourse as "a specific ensemble of ideas, concepts, and categorizations that are produced, reproduced, and transformed in a particular set of practices and through which meaning is given to physical and social realities" (1995, 44).

For Weale (1993), a belief system is composed of a set of assumptions. For Hajer (1995) and Dryzek (1997), in turn, a discourse is composed of storylines.[6] The latter are narrative constructions about social and environmental reality, enabling distinct elements of these domains to be combined in such a way as to make it possible for different social actors to reach a common understanding in their view of these areas (Hajer, 1995, 62). But what would be the storyline of EM? Weale (1992) argues that not only is there no canonical statement of the EM discourse, as it is also a multifaceted one. Therefore, it should not be seen as a coherent ideology whose elements are well articulated and in which there is substantial consensus about its meaning. According to Weale (1992), EM is permeated by some central propositions, all of which are subject to better intellectual elaboration. The emphasis and importance given to each of these propositions will then produce different styles of criticism with completely different political consequences. Hajer makes similar considerations. In his view, EM presents not one but a series of storylines. The discourse on EM is based "on some credible and attractive story-lines: the regulation of the environmental problem appears as a positive-sum game; pollution is a matter of inefficiency; nature has a balance that should be respected; anticipation is better than cure; and sustainable development is the alternative to the previous path of defiling growth" (Hajer 1995, 65).

If the EM discourse is formed by several propositions, it seems difficult to present a singular and general view of it. However, although it has such characteristics, this does not prevent us from grasping its format, as these authors try to illustrate. The first step in doing so is to contrast the discourse of EM with some of the assumptions inherent in the environmental politics of the 1970s. For Weale (1992), the status of EM as an ideology is largely based on the rejection of the validity of these assumptions that held through the 1970s. As Weale writes, the "structure of ecological modernization as an ideology is given by the denial of the general validity of these assumptions" (75–76). Hajer (1995), making a similar argument, contends that EM brought dramatic changes in the way European environmental policy was conceptualized, and this rupture, he argues, took place on the basis of the same assumptions that Weale (1992) notes. Boland (1994) provides an overview of the changes that the EM discourse has brought about in Europe through a literature review. We will use this author's work to clarify what these differences are. This will further clarify the assumptions that Weale (1992) and Hajer (1995) refer to when they claim EM would be guided by them.[7]

Until the 1970s, the environmental policy paradigm assumed the notion of a zero-sum game between the costs of environmental protection and economic growth. The EM discourse offered a new perspective on this point: It began to consider the possibility of a positive-sum game between economics and environmental protection. An example of this is environmental efficiency achieved through technological innovations that help reduce the cost of the production process by increasing its environmental efficiency. In contrast, the paradigm of environmental policy that prevailed until the 1970s did not recognize the interdependence between economics and ecology. With the discourse of EM, this interdependence is recognized as a prerequisite for the stability of economic growth and for the legitimacy of the state itself. The paradigm that prevailed until the 1970s also expressed great confidence in scientific knowledge. EM began to recognize the limitations of scientific knowledge in guiding policy decisions and the need to incorporate the precautionary principle into the environmental decision-making process. The approach that prevailed in environmental policy until the 1970s was also characterized by a fragmented view of environmental problems. Since the 1980s, regarding the EM discourse, these approaches have become more integrated and systemic. In the context of the old paradigm of environmental policy, there was also an insularity of the environmental decision-making process that was vulnerable to the strong influence of economic interest groups while remaining inaccessible to environmental groups. In the discourse of ME, there are increasing efforts to include environmental groups in the decision-making process. Apart from these differences, environmental policy in the old paradigm was guided almost exclusively by regulatory approaches based on uniform emissions standards. With the advent of the new paradigm of EM, experiments with more economic approaches began. The old paradigm of environmental policy also lacked a framework for monitoring and evaluating environmental policy itself. With the paradigm shift brought about by EM, the implementation deficit was recognized and efforts were made to overcome it. Finally, the two paradigms differ in the scope of environmental policy. Until the 1970s, environmental policy was strongly national in scope. Since the discourse of EM, greater emphasis has been placed on the establishment of environmental regimes and the relationships between national and global environmental policy.

All these distinctions are important here because, according to Weale (1992) and Hajer (1995), the ideological or discursive dimension of EM emerged from these ruptures that developed within European environmental politics. In this view, EM would have appeared as a discourse in the corridors of European environmental policy, which would have embarked on a revision of its principles because of its recognized limitations. It would thus be possible to view EM as a kind of institutional learning in which its principles emerged from a critique of the assumptions that guided environmental policy until the 1970s.

Another important point about these differences between the old environmental policy paradigm and EM is that these differences show some coherence in the EM discourse. The defining features of EM can be sought in these differences between EM and the environmental policies that existed in Europe previously. At the same time, and this is very important for our discussion here, the EM discourse has a core proposition without which it would lose much of its appeal. Among the various ruptures the discourse of EM has produced in European environmental policy, then, this is the most crucial. For Weale (1992, 76), the most important rupture of EM lies in the reconceptualization of the relationship between the economy and the environment. To Weale (1993), "the central proposition of ecological modernisation, as it developed in policy documents, was contained in the claim that environmental protection should not be regarded as a burden upon the economy but as a precondition for future sustainable growth" (207).

Dryzek (1997, 143) also sees the idea of transforming the capitalist economy to reconcile economic development and environmental protection as the central storyline of the EM. And in a similar vein, Hajer (1996) argues that EM is an environmental policy approach that starts from the "fundamental assumption that economic growth and the resolution of ecological problems can, in principle, be reconciled" (26). This seems to be the most important postulate of EM because, as these authors suggest, the other assumptions depend on this more general statement, at least as far as the possibility of giving coherence to modernization as a discourse or ideology of European environmental policy is concerned. We can conclude that the central storyline of the EM discourse is based on the idea that economic growth and environmental protection are compatible. It is a discourse that sees the environmental crisis as the result of the failure of institutions in modern societies but believes that the reformulation of these institutions can foster a process of environmental protection. Having made these reflections on the EM discourse, we will now analyze the condition of the EM as environmental sociology.

ECOLOGICAL MODERNIZATION AS
ENVIRONMENTAL SOCIOLOGY

According to Mol (1995), from the late 1980s onward, environmental sociology experienced a renaissance in which its themes and concepts underwent significant changes. Such changes fostered a mutual interpenetration between general sociology and environmental sociology, providing a fertile basis for creating a consistent sociological framework to analyze the contemporary ecological crisis. This mutual relationship between contemporary sociology and environmental sociology had some implications for the EM theory. One of them is that the conceptual framework of EM sociology is fundamentally inspired by the theories of modernization and postindustrial society. Regarding modernizing theories, the influence between these theories and the sociology of EM is explicitly recognized by Spaargaren (2000). For the theory of EM, Spaargaren argues that the environmental crisis is a vehicle for a deeper rationalization process in which new subsystems emerge to deal with environmental problems because existing institutions are not able to provide a good response. In examining the ecological crisis from this perspective of change, where EM is seen as the result of a rationalization process that leads to ecological rationality, Spaargaren reminds us that he and other theorists of EM are linked to theories of modernization in sociology. Because of this conceptual and theoretical connection, EM is influenced by the sociology of modernization as found in authors such as Weber and Parsons and in the more contemporary works of Habermas, Luhmann, and others.

As an ecological variant of modernization theories, EM turns to the process of "emancipation of ecology." This notion alludes to the growing independence of ecological rationality vis-à-vis other rationalities, specifically economic. The "emancipation of ecology" is, in this sense, a process of rationalization. As Leroy and Tantenhove (2000, 194) note, this basic idea of EM, of the growing independence of the ecological sphere or system, is closely related to the classical sociological understanding of the modernization process. Among the many ideas that made up modernization theory in the 1960s and its various contemporary reformulations is the general assumption that change in social systems can be understood as a process of structural differentiation and functional specialization.[8]

Joseph Huber (2000), considered one of the pioneers of EM theory, is appointed by Mol (1995) as the author most closely associated with modernizing and contemporary systemic theories. While acknowledging the influence of these theories in his work, Mol, on the other hand, seeks to bring the EM theory closer to authors such as Giddens, who, in many ways, may be seen as a critic of these traditions of sociological thought. In any case, EM theorists like Mol continue to nurture sympathy for systemic and modernizing theories as they seek to keep the general formulation provided by the founding father of EM (Huber) intact. Thus, in *The Refinement of Production*, Mol (1995) reminds us that his use of Huber's concept of EM does not occur without modifications and adjustments in his own understanding of the concept. But at the same time, he makes clear that these modifications are made in such a way that the core idea remains untouched, "leaving the general idea and concept unchallenged" (41).

Modernization theories see social change as a process of functional differentiation and rationalization. The former refers to the emergence of functional arenas of social interactions, and the latter to differentiated forms of rationality within these arenas or subsystems (Seippel, 2000). For a theorist of EM such as Mol (1995), this process of functional differentiation takes the form of a disembedding process. Following Giddens (1990), he argues that with the advent of modernity, social relations were disembedded from their traditional contexts and reembedded in new patterns, in terms of temporal and spatial organization. Just as the emergence of capitalism implied the emergence of an independent economic sphere, a similar process occurred with ecology.[9]

The affinity of EM to theories of postindustrial society is recognized by Spaargaren and Mol (1992). In their view, by examining the relationship between the industrial system and its ecological foundations, EM theory is placed in the realm of theories of industrial society. In this way, EM theory, in their opinion, "highlights the industrial rather than the capitalist character of modern society" (336). This association is often affirmed by supporters of this approach. For example, in *The Refinement of Production*, Mol (1995) points out that his study focuses on the industrial dimension of modern societies.

EM, Mol notes, should be interpreted as a theory that seeks to examine the restructuring of the industrial system in order to provide an answer to the colonization of nature by technical rationality (1995, 38). But what would this emphasis on the industrial rather than on the capitalist character of modernity mean? Spaargaren and Mol (1992) answer the question based on the dilemmas that permeated environmental sociology from the 1980s onward.[10] According to them, environmental sociology, influenced by contemporary sociology, was divided due to a dispute over the centrality of the dimensions of industrialism or capitalism in explaining modern environmental degradation. Thus, according to Mol (1995), both a (neo-)Marxist strand and a (post)industrialist strand emerged in environmental sociology to analyze the ecological crisis. Throughout this dispute, Spaargaren and Mol (1992) place EM in the series of theories that define modernity from its industrial dimension.

The neo-Marxist approach, in its analysis of the origin of the environmental crisis, tends to see capitalism as the main institutional feature of modernity. From this perspective, the accumulation process of capitalist (mega) corporations are the main cause of the environmental crisis. The second perspective, the postindustrial one, is divided into two sides. In it, there is what Mol (1995) calls theorists of hyperindustrialization (superindustrialization) and also advocates of demodernization (or deindustrialization). Against these last two perspectives, the neo-Marxist approach, by emphasizing the relation between production and class conflict, minimizes the importance of the forces of production in creating the environmental crisis. However, in the postindustrial perspective, the technological and industrial profile of the capitalist production system, not the process of capitalist accumulation, is the problematic aspect. An indication of this is the large-scale environmental degradation that has taken place in countries of "actually existing socialism." Therefore, class conflict and capitalist logic, as a whole, can be seen as important for understanding environmental degradation in some countries, but not as determining factors for understanding environmental degradation in modern societies in general, specifically in countries where the economic order was not conducted by the guidelines of economic liberalism.

We will return to these questions later. Having made this preliminary presentation of EM, we will now evaluate a central aspect of EM that permeates both the discursive and the sociological aspects and which, in our view, can define its contribution to contemporary environmental sociology.

INTEGRATION BETWEEN ECONOMY AND ENVIRONMENT

Many authors draw attention to the fact that EM has a strong emphasis on the economic aspect. After all, it seeks both "ecologizing the economy" and "economizing the ecology" (Mol, 1995). Weale (1992; 1993), Hajer (1995), and Dryzek (1997) agree that the central storyline of the EM discourse lies in its economic dimension. It is in its redefinition of the relationship between economy and the environment that lies its most decisive break with the propositions of environmental policy from the 1970s. Both the discourse and sociology of EM seem to share the same assumption: that it is possible to reconcile economic growth with environmental protection. Thus, what is behind the ideas of "emancipation from ecology" and "ecological restructuring of the industrial society" is that economic growth and environmental protection can go hand in hand. Next, we will take a closer look at this common argument. For this, it is necessary to revisit several issues related to the theme of contemporary economics and its relationship with environmental degradation.

We will start with Jacobs's (1991) work *Green Economy*, which presents the main problematic points involving the relationship between the contemporary economy and ecological degradation, seeking to delineate from there what could become an ecological economy. The discussions proposed below are based in large part on the work of this author. Subsequently, we will confront the interpretation by Jacobs with that provided by representatives of EM theory.

Two key points are part of the relationship between market forces and environmental degradation. One of them is the mechanism used to allocate resources: the use of individual decisions to achieve collective results. The second refers to the tendency in modern life for market forces to stimulate the physical growth of production. It is the combination of these two aspects inherent to market forces that define, according to Jacobs (1991), the interfaces between the economic system and the environmental issue. Market forces prevail when most decisions made by individual consumers and firms take place in contexts where no one in particular determines the collective consequences. Thus, market forces are put in motion from millions of decisions made by individuals or firms that are separated from each other. So in the capitalist market system, the allocation of resources (energy, raw materials, labor power, waste) tends to be decided not by a central agency (e.g., the state), but by decisions of individual agents (consumers/producers) who operate in the market.

However, there are several reasons to believe market mechanisms can fail to prevent environmental degradation. There are no guarantees, for example, that the market will promote the environmental protection of common resources. This is because its mechanisms (supply/demand) do not necessarily avoid consumption and, consequently, the exhaustion of certain resources. Scarcity may be the reason for an increase in the prices of certain resources, but this does not guarantee that degradation will cease. Scarce goods can be part of a more restricted market and become highly desirable for consumers with high buying power.

Another problem lies in the fact that the targets of environmental protection are public goods. They are indivisible and not subject to distribution to consumers. For goods of this type, it is not possible to discern between consumers and nonconsumers and therefore between paying and not paying for the good or service being offered. This implies that if goods of this type were supplied by the market, all people would become potential consumers, without restriction. This universal and collective characteristic of public goods makes it impossible to attempt to use only the logic of the capitalist market to promote the production and supply of goods and services that express these characteristics.

As for economic growth, the contemporary market economy tends to encourage the constant expansion of production, causing resources and ecosystems to be explored beyond their limits. However, the antagonism between the defenders of economic growth per se on the one hand and its detractors on the other is immersed, as Jacobs (1991) points out, in some misunderstandings. Environmentalists are right to be concerned about the physical dimension involved with economic growth, as there are limits to ecological systems that, regardless of whether they can be expanded or not, must be respected. In this sense, environmentalists are right to criticize economists who do not consider the limits of the capacity of the environment to provide resources and assimilate the waste produced. However, they end up making two mistakes when they associate economic growth directly with increased biophysical productivity. They overlook, in the first place, the fact that economic growth has several definitions. Economic growth, as the increase in biophysical productivity, as Jacobs indicates, is just one of them. Second, they are wrong in concluding that because current patterns of growth are dangerous the solution is no growth.

There are essentially three different definitions of economic growth: (a) increase in the biophysical output of the economy, (b) income growth (measured by GNP), and (c) increase in wealth (Ekins and Jacobs, 1995). Environmentalists and economists tend to use different definitions. While environmentalists often refer to the first meaning (biophysical production), economists seem to refer to the second (increasing gross national product [GNP]). Thus, the definitions do not represent the same thing. Although growth in GNP may coincide with growth in biophysical production, the relationship between one and the other is not direct, nor does it necessarily need to occur. As Jacobs (1991) makes clear, economic growth can take different forms. In some cases it is accompanied by an increase in resource use and pollutant emissions, but in other cases it is not. The central issue underlying the relationship between economic growth and environmental protection is physical or material production. According to Jacobs, firms need to maintain or expand profits, but this need not always be done through greater use of natural resources or through greater pollution. Theoretically, he says, one can assume that financial growth is possible even if there are environmental constraints (Jacobs 1991, 26).[11]

This indicates that it is possible to ensure that the increase in GDP does not lead to an increase in the use of resources and the generation of pollution, as long as the content of the growth is altered to incorporate economic activities with a decreasing environmental impact, at least when the value added to production does not necessarily imply a subsequent increase in the biophysical production of the economy. This is the path to greener GDP growth without increasing biophysical production. One of the first steps to achieve this goal would be the creation of an environmental impact coefficient (EIC). This coefficient is defined by Jacobs as the "degree of impact (or amount of 'environmental consumption') caused by an increase of one unit of national income" (1991, 54).

One operation the EIC requires is the measurement of consumption and environmental capacity. The latter is nothing more than the potential of the environment to perform its different functions (resources, waste assimilation, environmental services). Environmental consumption, on the other hand, corresponds to the use of these functions without necessarily disrespecting their maintenance capacity (Jacobs 1991, 73). The EIC would allow us to know which units of the GNP are growing and which are exercising direct pressure on the environment. It would thus allow separating the sectors of the economy responsible for greater pressure on the environment from those that exert less pressure.

For the coefficient of environmental impact to be lowered, a change in the cause-and-effect relationships involving economic growth and environmental impact is required. Between these two points, there are a number of transformation processes that, as Jacobs (1991) suggests, can mitigate or exacerbate the conflict between economic growth and environmental impact. For example, for economic growth to have an impact on the environment, income must be transformed into X amount of energy consumption. This amount of energy, in turn, requires Y amount of fuel consumption, which ultimately means Z emission of pollution. The relationship between these quantities is not fixed. The final environmental impact of X, Y, and Z quantities may vary depending on the energy source, type of product, type of consumption, production process, and treatment of industrial waste. Therefore, greater efficiency in these relationships is possible and desirable. The environmental impact coefficient should be composed of three different types of efficiency coefficients that, once modified, could drastically reduce the environmental impact of economic growth. These coefficients relate to (a) the production required by a unit of GNP, (b) the amount of environmental output required by a unit of production, and (c) the degree of environmental impact caused by unit of environmental output. According to Jacobs (1991, 104), if it were possible to increase environmental efficiency in all these dimensions, it would be possible to reduce the coefficient of environmental impact coefficient.

For this environmental efficiency to be achieved, a structural change in the economy is necessary to keep the environmental impacts within the limits required by the environment. This change would span the entire economic cycle and should address six central topics: (1) renewable resources, (2) nonrenewable resources, (3) pollution reduction, (4) the production process, (5) products, and (6) consumption.[12] Then, the possibilities of reducing the environmental impact of economic growth would be possible. Once incorporated into the economic process, they could significantly reduce the impact of economic growth on the environment. One of the central problems is that certain existing technological advances are not widespread and widely used. In this case, the current challenge would be not only technological but political. From this framework, the following question arises: How should environmental policies proceed to encourage technological innovation in the economy? The question here is to know how economic activities can be influenced to reach consensually established environmental goals.

The state can play an important role in this process. Jacobs (1991) points out four different types of mechanisms that could be used by the state to achieve environmental goals: (a) voluntary mechanisms, (b) command and control regulation, (c) government expenditure, and (d) financial incentives. Voluntary mechanisms are all those actions that individuals, groups, and companies take to protect the environment but are not coerced by law or driven by financial incentives. Regulatory mechanisms, in turn, include every administrative measure taken by the government that has the support of the law but does not involve direct government spending or the use of financial incentives. As for government expenditure, it can take two different forms: subsidies and direct government actions. The first addresses private companies; the second, government agencies. Government expenditure takes the form of subsidies if actions aimed at protecting the environment are voluntary and performed by nongovernmental actors. It takes the form of direct government action if the initiative starts with the government itself. Financial incentives seek to make environmentally hazardous activities less attractive in economic terms, making them more costly. At the same time, they seek to reward the most sustainable activities. Financial incentives then use the price system to achieve environmental goals, seeking to influence the economic decisions of producers and consumers.[13]

There are several important considerations to be made concerning these different environmental policy instruments. It should be noted that two out of the four constitute the main mechanisms for implementing environmental policy: command regulation and financial incentives. While voluntary mechanisms are generally welcome, they do not assure that the key players in environmental degradation will take any action. In this case, voluntary mechanisms are not substitutes but complements to instruments that imply more rigid control by the state. Government expenditures, in turn, place the costs of environmental protection on all members of a society and therefore make no distinction between those who are directly involved in the production of environmental problems and those who are not.

Regulatory mechanisms and market instruments, in turn, not only rely on the law and thus have greater enforcement power but are also based on the polluter pays principle; that is, they allow the costs of environmental protection to be imposed not on all citizens but on those who most obviously cause the environmental problems. As Jacobs (1991, 149) notes, unlike regulations and market mechanisms where the polluter of the environment has to bear the costs incurred, public expenditures spread these costs among all citizens. Therefore, we can assume that public spending, when applied, violates the polluter pays principle and therefore may shift the costs of dealing with pollution to those who have no direct connection to it. For these reasons, we can conclude that financial regulations and incentives tend to play a larger role in environmental policy. Having made this assessment of the triangle of economic growth, environmental protection, and policy regulation, let us examine these issues in the context of EM theory.

ECOLOGICAL MODERNIZATION AND
THE GREENING OF THE ECONOMY

Although the works of Mol and Spaargaren (1993) are important to situate the sociological orientation of EM more generally, the works of Jänicke et al. (1990; 2000) and Gouldson and Murphy (1998) are more useful for showing the ecological change in economic growth from the perspective of EM. In this part, I will use the work of Gouldson and Murphy in particular to explore this question.

The emancipation of ecology, to which EM theory alludes, encompasses two distinct processes that directly influence the relationship between economy and ecology: the process of "ecologizing the economy" and that of "economizing the ecology." Although this is not always stated, it is assumed that these two movements will end up resulting in economic growth that converges with the promotion of environmental sustainability. The first process involves a technological and organizational change in economic activities. On this point, EM means the replacement of curative technologies with preventive technologies. The latter are considered vital for delinking economic growth from resource inputs. Economizing the ecology implies, in turn, the introduction of economic concepts, mechanisms, and principles aimed at protecting the environment. This process takes place at two different scales: micro and macro.[14]

At the macro level, EM seeks to describe and promote a change in the structural character of the economy of more industrialized societies, causing resource-and energy-intensive industries to be replaced by knowledge-intensive industries oriented by the added value of goods and services. The economic structural change that EM seeks to encourage can be found in the work of Jänicke (1990).[15] In *State Failure*, he argues that in recent times, we can see two different trends affecting contemporary societies. On the one hand is the trend toward superindustrialization; on the other, postindustrialization. Superindustrialization is characterized, for Jänicke, as a traditional way of producing goods based on a quantitative growth of production. Postindustrialization, in turn, points to qualitative economic growth. In the latter, the basis of the economy is centered on the use of renewable resources, information, and products with high added value. In it, there is a preponderance of a nonmaterial type of production (information and services). In this scenario, the economy in general would prove to be more ecological, since it would be based on knowledge, information (nonmaterial goods), and the high added value of products (and no longer on the volume of material production).[16] Such a trajectory of the industrial system could promote *qualitative* growth based on the high value of what is produced.

EM theory, as defended by Jänicke and other EM theorists, presents a series of assumptions regarding the trends that accompany the emergence of the so-called information society. As noted by Jokinen, Malaska, and Kaivo-Oja (1998), theories of postindustrial society, such as the one by Jänicke (1990), assume that the production and exchange of information can displace the production of tangible goods as the primary focus of economic activities. The structural change in the economy should thus lead to the decline of the most polluting industries (manufacturing and agriculture) and encourage, as Jokinen, Malaska and Kaivo-Oja (1998, 493) observe, the dominance of nonpolluting industries.

At the micro level, EM establishes a central role for technological and organizational change at the company level (Gouldson and Murphy 1997).[17] For the EM theory, reconciling economic growth and environmental protection makes it necessary for the production system to incorporate environmental technologies. The latter are technologies that "reduce the absolute or relative impact of a process or product on the environment" (Gouldson and Murphy 1998, 29). However, they can be divided into control (or curative) technologies and clean (or preventive) technologies. The former do not change the production system as such but introduce additional technological systems (end-of-pipe technologies) that capture pollutant emissions to reduce their impact on the environment. Clean or preventive technologies, in turn, do not seek to treat pollution after its emission but to prevent or reduce such emissions in advance. Its focus is on the causes of environmental degradation, not the effects. Clean technologies are based on the principle of prevention, while control technologies are based on the principle of reaction. The greatest interest of EM theory is in promoting the incorporation of preventive technologies.[18]

Furthermore, technological change can happen incrementally or radically. Incremental change involves a gradual improvement of existing technologies and techniques. It seeks to improve existing systems and is based on curative technologies. More radical changes, in turn, encompass a discontinuous technological change involving the introduction of new technologies and techniques that break with the existing standard. From the EM perspective, it is not technology per se but a specific technological trajectory (preventive technologies) that can be seen as an ally to environmental protection. As Mol points out, according to EM theory "environmental technologies can no longer be labelled ineffective, nor can advanced technological developments be interpreted as all-devastating for nature" (1995, 44). For many ecological modernizers, such changes require some form of government intervention. And the success of this intervention depends on the capacity of the state to stimulate the technological transformations outlined above. These authors are skeptical about the possibilities of market forces leading to the changes needed to make the economy and environmental protection compatible. As Spaargaren points out, "ecological modernization is targeted primarily at market actors and the industrial sector," but "its main bearer should still be the state" (2000, 46).[19]

Among the different forms of regulation that EM inspires, there are rule-directed arrangements and rule-altering arrangements.[20] Gouldson and Murphy define them in a way very similar to that of Jacobs (1991) as we saw earlier: The first is perceived as a system of direct control over the organization and activities of the market operated by the government and its representatives, which has a legal basis and is operationalized through structures and implementation procedures. The second is defined as all those actions that are not imposed by law and that individuals, groups, and firms voluntarily exercise to protect the environment (Gouldson and Murphy 1998, 55). So, as they point out, one of the main characteristics of rule-altering, to the detriment of rule-directed, is that it seeks to promote environmental action without necessarily making use of the law, although there may be a threat to use it. Each of these regulatory forms has strengths and weaknesses, so the central issue involving these forms of regulation is not the replacement of one with the other but the possibility of complementing them in the political process.

It is also important to emphasize that the political experiences stimulated by EM are not restricted to these two types of mechanisms. For example, in rule-directed and rule-altering arrangements, financial intervention mechanisms can be used, although they are considered to involve a type of regulation distinct from the others.

The fact that financial incentives use a market instrument to promote environmental regulation leads them to be seen as a market-based solution to environmental problems. The opposite sometimes seems to be the case for those that attempt to guide environmental policy through regulatory tools supported by more direct government intervention. Thus it seems that the statism some authors see in the theory and discourse of EM (Barry 1999b) tends to place it on the left wing of the political spectrum, while the emphasis of EM on market mechanisms seems to place it more on the opposite side. Attempting to differentiate such instruments through this prism may be a mistake, however, as Jacobs (1991) points out, because both taxes and regulations operate in markets and are generally introduced by governments to influence economic decisions. Economic incentives, in this case, cannot be considered as originating in a "free market" because they are instruments of public policy implemented by governments. Thus, as Jacobs reminds us, both economic incentives and regulations aim to change "free" decisions that, without their influence, would lead to undesirable outcomes.

In this case, it makes no sense to see EM, at least in the versions presented by some of its representatives, as an approach based strictly on the free market. For the use of economic incentives alone, even if it were true (which it apparently is not) that EM is exclusively oriented toward them, requires regulatory intervention.[21] EM theory, at least for some of the works examined here, presupposes ecologically regulated capitalism. The state plays a central role in encouraging ecologically responsible behavior by producers and consumers. Likewise, to see an approach in EM with a strong statist bias, as Barry (1999b) does, is also a mistake. According to Hanf (1994), EM represents an alternative to these two trends invading environmental policy. EM strategy is based on a wide range of organizational actors (mainly the state and companies) that must regulate their actions to allow life in common. It is very unlikely that environmental policy inspired by EM will be based on just one of the regulatory instruments we referred to earlier; rather, it is likely that it will be a mixture of these different instruments.[22]

The main flaw of EM theory with respect to these issues is related to the little attention it seems to give to environmental regulation. As Gouldson and Murphy (2000, 35) note, although the literature on EM is quite extensive, there are few studies that focus on the role and importance of regulation in environmental policy. As a result, we can conclude that EM theory does not do justice to the variety of policy processes that can be used for government intervention in the relationship between business and environmental protection and the considerations that can arise from these interventions.

TOWARD A CRITIQUE OF ECOLOGICAL MODERNIZATION

One of the obvious areas of sociology into which the environmental issue can be incorporated is the sociology of industrialism. This is an already well-established field of research in the social sciences, where environmental issues could prove to be highly relevant (Martell 1994). EM theory seeks precisely to follow this advice. Mol (1995), one of the main proponents of this theory in the field of environmental sociology, places EM precisely within the wing of postindustrial society theories. For him, EM theory seeks to both theorize and also promote the institutionalization of ecology in the production and consumption processes. One of the strengths of EM lies in the fact that it thus contributes to studies that seek to show that it is possible to go beyond the conflicting relationship that exists between the economy and the environment. However, this does not free EM from contradictions, as we will see below.

Much of what EM theorists produce is in line with what some authors establish as necessary to enable ecological economic growth (Jacobs 1991; Ekins and Jacobs 1995). This is also because industrialism and capitalism have affinities with each other. One of them concerns the propensity that both have for technological innovation, which is of fundamental importance for EM. In this respect, it is even possible to imagine a rapprochement between EM and Marxist theorists. This approximation seems to be possible since the theme of economic growth, or the process of capital accumulation for Marxists, is part of the theoretical interest of these two traditions of sociological thought. As Raymond Aron wrote in *Lectures on Industrial Society*:

> The sociological problem which has provided the main theme of this book [*18 Lectures on Industrial Society*] is that posed by Marx and Marxism—especially as the latter is expounded in Capital. . . . [Marx] tried to understand the laws of its development . . . the central phenomenon in Marx's view was that of accumulation. He believed that the essence of capitalism was to be found in the accumulation of capital. By choosing economic growth as the central subject of this investigation I have taken up the Marxist theme of accumulation in the terminology and using the concepts of modern economics. . . . Instead of capitalism I have chosen industrial society (or technical, scientific or rationalized society) as the principal historical concept. (Aron, in Bell 1999, 73)

For this reason, Bell concludes, in *The Coming of the Post-Industrial Society*, that authors like Aron, who prefer to emphasize the idea of industrial society rather than the concept of capitalist society, "pay homage to Marx by presenting the forces of production as the central idea" (1999, 73). This seems to indicate that under certain aspects of the analysis of modern society, Marxist and postindustrial theorists are closer than we might think. At the same time, this allows us to say that because of these affinities, the environmentalist critique of industrialism can also be directed toward capitalism, at least when it addresses the environmental consequences generated by the "productive forces."

As we have seen, it is a mistake to try to equate economic growth directly with environmental impact. No growth rate in itself—negative, stationary, or positive—can tell us, as Jacobs (1991) reminds us, what is happening to the environment. None of these goals in themselves can be seen as a useful target for ecological policy. The challenge seems to lie in knowing under what conditions economic growth can become environmentally benign or dangerous. Different types of economic growth can have very different impacts on the environment. Thus, Ekins and Jacobs (1995) argue that ecological economic growth is theoretically and practically possible. The problem seems to be in establishing the necessary changes and how to put them into practice. First, it is necessary to encourage a process of environmental efficiency throughout the economic cycle, involving both the extraction of resources and the dumping of products into the biosphere as well as production and consumption patterns. As these authors observe, many technological and organizational innovations already prove to be accessible to enable greater environmental efficiency in economic activities. Second, the great challenge is to create policies that allow the use of these techniques and innovations in the real world.

Whether the decoupling, in terms of environmental inputs from economic growth, can be absolute and effective in the long run is perhaps questionable in some respects. It is evident that many of the measures that seek the efficiency of the economic system can reduce and delay the impact of economic activities on the environment. But more long-term problems can also occur (an entropic problem). Obstacles must also be considered when trying to implement measures aimed at achieving this environmental efficiency and encouraging of technical-scientific development.

EM theory has made several contributions to this debate. It does this by focusing on the macro and microeconomic aspects of change and, simultaneously, on the capacity for government intervention. Several of the EM proposals are very close to the green economy suggested by Jacobs (1991). For example, at the macro-structural level, many countries appear to be already going through the postindustrial trajectory that Jänicke has pointed out in *Failure State*.[23] In a study conducted by the author analyzing thirty-one industrialized countries, he has noticed the emergence of a free environmental effect arising from the structural change that took place in these economies. In countries where there was an improvement in the relationship between economic growth and environmental impact, this improvement was seen as a result associated with a change in the economic structure of these countries.[24] These economies are no longer composed essentially of energy-intensive industries and material resources and began to be characterized, for the most part, by knowledge-and service-intensive industries.

But not all theorists of EM focus on this structural change. One of the problems associated with it is that, as Gouldson and Murphy (1998) and Simonis (1985; 1989) note, it has been involuntary. But such studies are not unimportant because they show that countries with similar economies and economic performance can have very different environmental outcomes, and knowing why this is so seems important for environmental sociology. So, as Gouldson and Murphy (1998, 2) remind us, while it is not clear that a complete and absolute synergy can be established between the economy and the environment, it is clear that there are ways to make this relationship less ecologically negative than it is today. Although there is no evidence that this "environmental gratis effect" is emerging as a dominant and general trend, it is certainly a strategic element of EM, at least in situations where this trend not only seems to be developing but can somehow be encouraged by public policies.

Jacobs (1999b) makes some observations that confirm many of these assumptions of EM theory. If recent globalization is bringing large-scale environmental degradation, he believes it is also leading to countervailing trends. With current globalization, production is becoming less intensive in material terms. The reason for this is that as companies become more globally competitive, they must raise their efficiency standards in production, which will encourage more efficient use of energy and resources. Demand is also changing, he notes. Many products are becoming smaller, and their value is associated with their aesthetic qualities rather than their materials. And many sectors related to the advent of globalization are associated with the provision of services (telecommunications, tourism, leisure, etc.) rather than material goods.

However, since structural changes such as those described by Jänicke seem to have occurred unintentionally, much of the EM literature seeks to delineate an environmentally possible technological trajectory based on a smooth intervention by the state. Thus, EM merges a strategic vision concerning contemporary trends with a prescriptive and interventionist vision based on the state. This change of direction implies a differentiated focus from many analyses, seeking to delineate, firstly, the innovations that capitalist companies must implement and, secondly, more specific analyses on how state intervention may proceed to encourage this process. The success of environmental regulation encouraged by EM will be determined by the level and nature of technologies and techniques that are developed and applied in response to it (Gouldson and Murphy, 1998). This will depend on the development of clean technologies and how much these technologies will allow the reduction of the environmental impact of economic growth. So the success of EM will depend on the possibility of gradually reducing the impact of economic growth on the environment without generating increasing costs, either for the state or for companies.

Some of the positive effects of policy intervention inspired by EM are already emerging. The most industrialized countries that have achieved a substantial decrease in their environmental impact, at least concerning some important environmental indicators, are those that created policies inspired by the EM discourse. Among them are Germany, Japan, the Netherlands, Norway, and Sweden. All these countries have managed to increase their environmental efficiency in the following dimensions: (a) high energy efficiency (in terms of the sum of energy required to produce a unit of national income), (b) decrease in per capita emissions of pollutants such as carbon dioxide and sulfur dioxide, and (c) decrease in per capita generation of household waste and other solid waste. Thus, not only would a structural change, prescribed by Jänicke and colleagues (1989), bring an "environmental gratis effect," but the political intervention in EM, in some countries, has already yielded concrete results in reducing the environmental impact of economic activities.

The way in which EM has examined the relationship between the contemporary economy and environmental sustainability can no longer be assessed in the same way as two or three decades ago, when the environmental movement advocated the need for "zero growth." Therefore, as much as there are doubts and uncertainties involving EM theory and discourse, it has brought new issues that cannot be ignored. Murphy's (2000) arguments on this subject can be seen as appropriate. Although EM theory does not provide a definitive answer to today's environmental problems, he says, the work in this area does provide valuable and important ways to think about environmental policy in the short and medium term. Its main contribution is that it offers alternatives to the existing conflict between the economy and the environment. Of particular importance here are the studies that indicate the progress that certain environmental policies in the richer industrialized countries have achieved in recent years. This, as Murphy notes, is an almost unique contribution in the literature of the social sciences, which tends to point to the inability of government and industry to deal with environmental problems. Therefore, we can add that while it is right to study the existing contradictions between the economy and the environment, the works that try to show situations in which this contradiction is mitigated or even ceases to exist are no less valuable. Concerning other environmental discourses that exist today, EM offers a more plausible strategy to transform industrial society into a more environmentally careful society, although it still does so in a capitalist way.

However, there are still problems with this approach. In a sense, the theorists of EM are caught in a contradiction, for EM seems to be fraught with all the shortcomings of approaches that try to argue that the capitalist market system is not linked to contemporary environmental degradation or, if it is, that it seems residual to understanding environmental problems. What is important for understanding the current ecological crisis is industrialism, not capitalism. The idea that environmental degradation is an endemic aspect of the capitalist market system is often challenged in two ways. The first points to the possibility of ecological capitalism and the second argues that environmental degradation is related not to market forces but to industrialism. The idea that capitalist market forces are not directly related to environmental degradation uses precisely these two arguments to make us believe that capitalism will become ecological by itself.[25] In the latter case, technological determinism is replaced by economic optimism, but the two are intertwined in some way.

It is precisely these two arguments that EM supporters use as a guideline to delineate the theoretical position of EM. Spaargaren and Mol (1992) argue that EM theory focuses on the industrial and not on the capitalist dimension of modernity. The consensual adherence of EM authors to the industrial dimension raises several important questions. Does this mean that, for EM advocates, contemporary environmental problems are associated only with contemporary industrialism and not with capitalism? Mol (1995) helps to clarify this point, claiming that ecological modernizers contest the neo-Marxist argument that ecological protection goes in a direction contrary to the logic of capital. For ecological modernizers, this neo-Marxist proposition would be true under the conditions of "jungle capitalism," but not in the context of a regulated capitalism. Furthermore, they argue that capitalist companies are increasingly capable of adapting to the requirements of environmental sustainability without the need to deprive themselves of favorable production conditions and new markets and their profits. Environmental protection even proves to be a profitable market for eco-industry (environmental technologies, management systems, etc.). The emancipation of ecological rationality, which authors like Mol and others take as given, is a process consistent with the rationality of capitalism itself. Therefore, green capitalism, so criticized by eco-Marxism and eco-anarchism, is seen as possible and even desirable by authors such as Mol (1995, 42).

Mol (1995) presents some additional arguments woven by ecological modernizers that lead us to questions of a more properly sociological nature. Ecological modernizers, relying on the theory of postindustrial society, believe that the very word "capitalism" is no longer adequate. This is for two main reasons: first, because capitalism has become an increasingly less distinguishable feature of Western industrial societies compared to other non-Western societies and is thus taking on a universal character around the world. Furthermore, no viable and possible alternative that could challenge capitalism as an economic system has yet been sketched.

Spaargaren and Mol (1992) and other theorists of EM seem to fundamentally overlook the dimension of capitalism. To them, capitalism is irrelevant because it is becoming, or will become in the future, sustainable. But this view has its pitfalls. The defense of ecological capitalism serves precisely to discredit the notion that market forces are involved in creating today's environmental impact.[26] Many of these arguments, however, are fallacious and based on a kind of reverse economic utopianism. By treating industrialism as the main axis of analysis of the ecological crisis, ecological modernizers run the risk of turning the concept of industrialism, and by extension the concept of EM, into a theoretical construct that obscures rather than illuminates understanding of the relationship between capitalism and contemporary environmental degradation.

The emphasis of EM theorists on the axis of industrialism can be understood only in the conceptual dispute that these authors seek to wage with neo-Marxist authors. However, there is no reason to privilege one institutional dimension of modernity (industrialism) over another (capitalism). As Goldblatt (1996) indicates, both dimensions of modernity are key variables for understanding modern environmental degradation. Industrialism could not, for this reason, be considered the exclusive cause of contemporary environmental change. It is, above all, a direct cause of environmental degradation and must be, according Goldblatt, "conceptualized in relation to other aspects of the social order" (38).

In a sense, EM theorists agree with the neo-Marxist theorists that capitalism generates environmental degradation. What the neo-Marxists declare, Mol (1995) asserts, may be true under conditions of "jungle capitalism," but would not hold under the regulated capitalism that has developed since World War II. Thus, the disagreement between neo-Marxists and ecological modernizers is not about capitalism per se, but regulated capitalism. For ecological neo-Marxists, capitalism, regulated or not, remains an important and worrisome cause of environmental degradation. For EM theorists, the argument is valid only for unregulated capitalism. The fact that neo-Marxists and ecological modernizers diverge on the issue of which type of capitalism (jungle or regulated) is environmentally damaging does not explain why industrialism is the main focus of EM. This makes their arguments on these issues somewhat confusing. The main question is what leads to environmental degradation under capitalism and how this relates to the clash between industrialism and capitalism within EM theory. Goldblatt (1996) provides an answer that brings us back to the questions already discussed in this chapter. He argues that at the heart of environmental degradation today is the problem of economic growth. However, Goldblatt emphasizes that the modern world's tendency and ability "for growth is a consequence of the dynamics of capitalism and not industrialism" (39). As Goldblatt observes,

> As Giddens rightly argued in his analytical assessment of capitalism, it is the investment-profit cycle in the context of competitive markets that is the institutional mainspring of the individual and corporate drive for expanded production and profit—in a word, for growth. That is not to say that other institutional frameworks cannot also unleash industrialism, although as state socialism has demonstrated with such clarity, none has yet managed to do so with the same degree of efficiency and effectiveness. (39)

The contradiction that authors such as Mol and Spaargaren (2000) produce in their defense of EM theory becomes evident. If it is the phenomenon of economic growth that EM theorists seek to reconcile with environmental protection, this is an aspect that, ironically, is linked more to capitalism than to industrialism. EM theorists could justify that economic growth is not an exclusive feature of capitalism but is also present in industrialism and (post) industrial societies. This argument has been incorporated by postindustrial society theorists since the 1970s on the grounds that economic growth could be seen as a central goal for both capitalist and socialist countries.[27]

However, economic growth in Western countries occurs on a totally differ- ent basis than in countries where economic growth is tied to state planning. In capitalist societies, economic growth is driven through competitive markets, not state planning. It occurs within a culture that seeks to promote economic liberalism. And the fact that state ownership of the economy is disappearing in many countries, as Jacobs (1991) warns, does not detract from the need to make such a historical distinction. These reflections are important for the debate itself about the fundamental characteristics of capitalism and refer us to the peculiar social values involved in the effort to promote economic growth in capitalist societies, on the one hand, and in societies of "actually existing socialism," on the other. In the context of capitalist societies, says Jacobs (1991), it is the dominant belief in the efficiency and freedom of market forces, which is part of the Western value system, that is linked to the continued physical expansion of the economy. Totally distinct are the values associated with economic growth that are shared by communism and collectivism.

Jacobs's argument that industrialism can become an ideological construct that overshadows the relationship between market forces and environmental degradation can thus be applied to the case of EM. If there are different mech- anisms by which economic systems produce environmental degradation, then uniting them in a "single ideological construct such as 'industrialism' does not help in understanding how the degradation occurs and how to prevent it" (Jacobs, 1991, 48–49). This, Jacobs continues, does not help us understand the ways in which capitalism itself tends to produce environmental destruc- tion. The problem with this is that when one emphasizes industrialism rather than capitalism in understanding environmental degradation, as EM does, one tends to view these beliefs and values as inherently virtuous for the environ- ment. From this view offered by ME, then the problem lies in technology, not in the values that govern market forces. In this case, economic liberal- ism itself. Or, in Jacobs's words, "So long as industrialised societies persist in seeing the beneficent invisible hand rather than the destructive invisible elbow, they will find it difficult to protect the environment" (1991, 48–49). The fact that ecological modernizers claim to be more concerned with indus- trialism than with capitalism is precisely because they do not see these values as problematic. They take them as given without having to change them; or these values become problematic only when they promote a particular path for industry and technology. However, the values associated with freedom from market forces relate not only to the ability to promote new technologies but also to the functioning of the economy as a whole, and they create the context in which the industrial system operates. Thus, contrary to what this view seems to imply, it is not clear that economic liberalism does not itself require some reconfiguration to adapt to the environmental agenda.

EM theorists make the same mistake they attribute to supporters of the concept of SD, although it is not obvious to us that the latter concept really expresses this kind of problem. According to Spaargaren and Mol (1992), one of the problems involving the concept of SD, in addition to its conceptual vagueness, is its objective to integrate ecological quality with economic growth via industrialization. Before being considered as the main factors involved in environmental destruction, economic growth and technological development are perceived, from the perspective of SD, as the main "condition for sustaining the sustenance base, rather than as the main cause of environmental destruction" (333). This criticism of SD leads to an ironic picture. After all, it is not the concept of SD, that seems to us to integrate environmental quality with economic growth through industrialization; it is the very concept of EM, which assumes that economic growth is a defining characteristic of industrialism and not capitalism, as we could infer from the arguments of Spaargaren and Mol above. This leads theorists of EM to fall into the error they themselves associate with the concept of SD, leading them to view technological change as encompassing greater innovation, while economic change is less significant because it could be viewed from the perspective of business as usual. ME proposes a major ecological change of the industrial system suggesting that this is possible without any change to the values attached to economic liberalism. But it is reasonable to assume that this change also implies the emergence of some kind of ecological political liberalism.

As stated by Jacobs (1991), many technological innovations could currently be used to mitigate the environmental impact of economic growth; however, several factors and obstacles prevent their improvement and wider use. Knowing why environmental technologies are incorporated (or not) means understanding how the process of technological and organizational innovation occurs in capitalist companies, and this process involves, as Jacobs observes, the broader economic relations in which modern technology itself is inserted. New technologies are not created separately from a political and economic context; they need to be accepted and financed based on economic calculations and broader social acceptance. However, it has not been in the interest of economic actors to develop technologies that protect the environment. To understand this resistance, we cannot limit ourselves to the technological sphere, but we have to assess the economic context in which these technologies are developed. The process of technological innovation that EM theorists consider necessary for the changes they envision thus goes beyond factors specifically pertaining to industrialism; it implies the economic calculation that market agents recurrently make and the more general social obstacles (e.g., risk acceptance) that involve technological innovation.[28] Moreover, if the process of economic expansion that characterizes capitalism is to remain under the lens of our concern, the theory of EM would neglect an important element in the debate about the "economization of ecology," for if this economic expansion remains ecologically dangerous, EM risks not making an important contribution to this debate.

The massive adhesion of EM to the axis of modern industrialism also raises other problems. After all, the sociological critique addressed to the theories of modernization and postindustrial society is extensive in the social science literature.[29] EM theorists leave many gaps in the way they incorporate this literature. In the case of modernization theories, for example, they mix authors traditionally associated with this strand of sociological thought with others who are not directly associated with it. Thus, for example, they quote Giddens, an author who is critical of the idea of modernization and of the theories that seek to endorse such an idea.[30]

EM theorists seem to see as similar the processes of structural differentiation advocated by modernization theorists and the disembedding process to which Giddens (1990) refers. Understandably, Mol (1995) makes this approximation between the two concepts. Structural differentiation involves the idea of dissolution, disembedding, and reinstitutionalization of new subsystems of social interaction and new forms of rationality (Seippel, 2000). The "embedding and disembedding mechanisms" of social practices by Giddens (1990) also alludes to these aspects. But despite this similarity, it must be considered that they contain very different assumptions. Functionalist theory, to which modernizing theories are generally linked, are characterized for the most part by the imposition of "system needs" and, consequently, by the introduction of some type of functionalist explanation. Structural differentiation emerges and develops because the social system requires it. But these are precisely the two central aspects that Giddens seeks to purge from the social sciences.[31]

In *The Refinement of Production*, Mol affirms that EM envisions an ecological transformation of the industrialization process, a transformation that would secure the sustenance base of the industrial system (1995, 37). From the demands of environmental protection in the 1960s and 1970s, a structural differentiation of society began that now makes it possible that "sustenance base can be guaranteed."[32] If for Durkheim structural differentiation was an answer to "social needs," the structural differentiation of EM now occurs in order to ensure the "sustenance base of society." In both cases, structural differentiation becomes a question of survival of the social system. In this context, two questions need to be asked: (a) What are the limits of the "sustenance base" (environment) referred to? Is it limited to the environmental resources that certain companies or economic sectors use for their activities, or would the "sustenance base" in question be planetary? And (b) What are the limits of the society in which this structural differentiation operates? If the first question leads us recognize that this "sustenance base" is local, as EM would take place in the most industrialized countries, then it is worth asking whether the language used is appropriate given the need for global environmental governance. Local responses, no matter how positive and valuable they may be, may prove to be insufficient to address these problems. Given the challenges posed by climate change, it would be strange to claim that any society's support base is guaranteed. In short, there is a functionalist language here that simply seems outdated.

The second question leads us to similar problems when we consider that the understanding of societies as nation-states has become obsolete in the wake of economic globalization. On the one hand, given the global dimension of the division of labor and, of course, of capitalism itself, modernization in certain sectors or regions is very limited. Finally, given these two questions, what does "guaranteed" mean in this context? The question is not whether the changes advocated by ecological modernizers are positive or not; the question is whether the language and conceptual framework in which they are addressed are appropriate. To put it another way, there are good reasons to advocate the changes EM theorists point to, but this should be done, it seems, with less enthusiasm and, to some extent, in less deterministic language. The determinism and projected directionality of these changes are reminiscent of the vices of earlier theories of modernization, which the sociological literature has shown to be wrong. It is simply wrong to imagine emerging changes, metaphorically, like the awakening of a caterpillar. The biological metaphor prevents a more prudent sociological view of understanding environmental change for the reasons given here.

These questions also extend to the view EM theory casts on the emancipation of ecological rationality. A problem associated with the concept of "ecological rationality" embodied in this theory seems to incorporate a questionable precept about the relationship between social systems and adaptation. In a functionalist interpretation, rationality, implying a process of structural differentiation, tends to be seen as a process of better human adaptation to environmental conditions, a precept that seems questionable given the risks brought by new technologies and the division of labor on a global scale.

To better illustrate this point, let us consider Giddens's critique of Parsons's functionalism in *The Constitution of Society*. In this book, Giddens notes that Parsons's evolutionary view is based on the idea of adaptation, which he argues tends to suggest "the reduction of uncertainties about the vagaries of nature and the reduction of uncertainties in respect of future events" (1984, 271). In neither of these senses, Giddens argues, does adaptation seem to take place in the modern industrial societies analyzed by Parsons. Echoing the critique generated by contemporary environmental discourse, Giddens asserts that "increased control over the material environment, yielded by technological development or the manipulation of authoritative resources, is by no means the same as reduced uncertainty of outcomes" (271). For him, a farmer who promotes more efficient labor techniques may prove as vulnerable to climatic variation as a hunter. If we consider the possibility of reducing the unpredictability of the future, Giddens asks, who could say that the modern world, with the uncertainties associated with economics, politics, and nuclear weapons, is less uncertain than the world experienced by Paleolithic people?

There is a danger, then, that the concept of ecological rationality involves a kind of determinism in which increasing structural differentiation is seen as a gain in adaptability; that is, one could see it as in Parsons, where the structural differentiation represented by the idea of ecological rationality takes us to a larger and more complex level of adaptation. Thus, to the extent that EM theorists admittedly emulate Parsons's thinking, as they themselves acknowledge, it must be examined whether the concept of EM risks reproducing the same anthropocentric arrogance that was present in earlier Parsonsian sociology. So, if it is certain that functional differentiation will accompany the process of EM, what guarantees are there that this will necessarily mean a reduction in the uncertainties that these are the factors that condition human life in its relationship with the environment. Again, the point is not that we cannot hope certain economic and technological developments will be more "environmentally friendly" but whether we should consider this question in light of these views. The principle here is that functional differentiation necessarily leads to better adaptability. This helps us to understand the nature of the critique that York and Rosa make of EM:

> If EMT [ecological modernization theory] is making the second argument—that institutional modernization actually helps to reduce environmental problems and bring about a transition to sustainability—then it attracts the burden of looking beyond changes in the structure of institutions and actually demonstrating positive environmental outcomes stemming from modernization. It must provide a theoretical specification of the connections between institution building and environmental outcomes. . . . It must also demonstrate this if it is truly about ecological modernization rather than merely about institutional responses to environmental problems. (2003, 275)

This occurs, as we have indicated, because EM theorists tend to attribute a positive ecological effect to the process of structural differentiation almost automatically. Social differentiation leads to ecological rationality and is inherently virtuous from an ecological perspective. The social system becomes more "rational" in dealing with environmental problems. Ecological sustainability in this case is the result of a process of structural differentiation. Since structural differentiation occurs on the basis of environmental problems, the change involves a number of questionable assumptions. One of these is the assumption that environmental quality will be secured and that a social consensus will emerge as a result of this process. It is as if, in the style of functionalist sociology, this process produces an adaptive performance that is taken for granted, as in Parsons. Thus, modernization theories are often accompanied by a strong economic optimism and a kind of anthropocentric arrogance that relies too much on the possibility of human control over the environment.[33] Smith, looking at the structural differentiation in modernization theories, notes the view these theories offer of this change:

> Specialisation allows man to maximise his control of the environment. He does this by evolving more effective roles and units. At the same time, these roles and units become increasingly autonomous; they are freed from biological and ecological constraints. In this way, differentiation combines with efficiency; by its operation, personalities and social systems are enabled to develop themselves, and to adjust their elements to their surroundings. Differentiation, in other words, bestows that most desirable attribute on structures—flexibility. To differentiate is also to develop and mature by adjusting in a realistic manner to new situations; and this capacity for adjustment and flexibility is exactly what differentiation alone can confer. Social development is also social maturation, and hence beneficial. (1973, 18)

It is not possible to adopt this kind of vision of social change without including some presuppositions of the normative order because that the process of adaptation involves the increasing control of society over the external environment. An attribute included in the idea of structural differentiation in modernization theories states that "it increases man's autonomy and society's maturity pari passu with its capacity for maximising his control over nature, and therefore is a beneficial process" (Smith 1973, 18). The theory of EM does not seem to provide a satisfactory check on these assumptions. If "ecological rationality" is a reflection of a process of adaptation of the social system, this change would signal a kind of "ecological maturation" of modern society and could therefore be considered beneficial. But a question arises here: Should any process of differentiation involving environmental issues be considered beneficial? As we will see in the next chapter, the SR offers a different argument on this issue: In many circumstances, increasing specialization in business and science creates the conditions for the emergence and amplification of environmental risks, rather than the opposite.

The sociological reading offered about EM clarifies little—in fact, it tells us almost nothing—about how these assumptions are now being articulated by new ecological reading of the idea of modernization. Hence, some criticisms that are directed at the theory address these points that seem to us to be essential. McLaughlin (2021) observes, then, that EM theory is "conceptualized as a process of adaptive upgrading" where the institutions and structures of modern society operate to drive change, a process where these institutions, for a supporter of EM theory like Huber, "are continuously developed and upgraded to ever-higher degrees of complexity" (Huber 2009, 46). Says McLaughlin, "EM theorists equate better adapted with more advanced, a context-independent assumption that ignores the capacities of 'pre-modern' societies to identify and solve environmental problems . . . as well as the possibility of 'advanced' societies adapting through simplification" (2021, 185).

EM is composed of very diverse literature that does not present the internal consistency many would like to find in it. Thus, authors such as Seippel (2000) question the theoretical merit of EM. According to the author, there is a certain agreement in the social sciences that theory involves a description and a general statement about the relationship between two or more entities. What distinguishes good from weak theories is therefore their internal consistency and logic. In this case, taking the "specific requirements of what a theory is," Seippel says, "it does not seem justified to speak of ecological modernization as a theory" (2000, 300). Says McLaughlin, "EM theorists equate better adapted with more advanced, a context-independent assumption that ignores the capacities of 'pre-modern' societies to identify and solve environmental problems . . . as well as the possibility of 'advanced' societies adapting through simplification" (2021, 185).

It really does not seem to make much sense to present EM as a grand theory.[34] The strength of EM, however, lies neither in the degree of generality of its formulations nor in its internal coherence as a theory. Although it can be seen as a contestable concept, a feature we will address in the next chapter when analyzing the concept of SD, its importance lies in the fact that it provides us with arguments for the debate on the integration of economy and environment. Although authors such as Seippel (2000) are correct in their critique, they make the mistake of reducing the role and importance of social theory to its explanatory and generalizing capabilities. Sociology can also function as a discourse that, although it does not have the characteristics generally associated with a theory, has its specific meaning for the social sciences as a discourse.[35]

To conclude this chapter, let us briefly consider the theoretical directions EM theory might take. As we have seen, it aims to make ecological economic growth possible. To this end, we believe it should abandon its unanimous emphasis on the industrial axis of modernity and consider capitalism as an important topic of analysis, and that might mean breaking with the view that sees business as usual as necessary and desirable. The purpose of EM, as we have seen, is by no means meaningless. Indeed, it is not only defended by proponents of EM. Eckersley (2000), for example, notes that the goal of achieving ecological economic growth is also dear to the social democratic vision of the relation between the economy and the environment in authors such as Jacobs (1991) as well as Daly, Cobb, and Zarsky. EM theory can only gain by strengthening its arguments based on the social democratic vision that Eckersley seems to endorse.[36] In these authors' view, the regulated market economy, as proposed by an ecological social democratic vision, offers the best institutional structure to accommodate the ecological interest of environmental protection. But if EM must include the dimension of capitalism in its analysis, what theories would be closer to its goals? As Gouldson and Murphy (1998) attempt to show, EM has helped to foster a new eco-industrial regulation. It is surprising, then, that no work has attempted to link EM theory with theories of regulation, or even suggested that this is the path it should take.

O'Brien and Penna establish a relationship between European environmental policy and regulatory approaches.[37] According to them, among the various concepts that figure as central in regulatory approaches, two seem to touch directly on issues that are at the heart of environmental policy and, in our view, of EM theory: regimes of accumulation and modes of regulation. The first refers to the relationship between accumulation and consumption. The second, in turn, is the basis for guiding economic growth over time in a specific regime of accumulation (1997, 194). So theories of regulation seek to work with how a different paradigm of production and consumption (regime of accumulation) emerges and investigate how such a regime stimulates the emergence of a new type of accumulation (mode of regulation). Furthermore, theories of regulation seek to point out the conflicts and struggles of interest that arise to promote a new mode of accumulation. These conflicts do not disappear but remain latent in the capitalist market system.

For the regulatory approach, the struggles and conflicts raised by environmental policies are vital to opening up new forms of accumulation. It is not the task of this work to assess the implications that regulatory theories and approaches may have on EM theory in detail; we only want to indicate that regulatory approaches can bring both conceptual and methodological contributions to it. As stated by O'Brien and Penna, "In our view, the regulation approach offers both conceptual resources for and a methodological insight into the interpretation of contemporary social and political change" (1997, 195). This assertion may also be valid for EM theory. Regulatory approaches seek to show how different forms of accumulation regimes and modes of regulation can exist within the capitalist system, and the challenge of EM to promote a new ecological economic growth seems to imply just that. Thus, one of the main contributions these approaches provide for EM theory consists in the fact that they illuminate the conflictual character that the emergence of a new regime of accumulation engenders, as well as the conflicts involved in its long-term maintenance. Thus, questioning whether EM is linked to capitalism or not turns out to be more a semantic rather than a substantive debate.

CONCLUSION

As demonstrated throughout the chapter, EM theory has been criticized, above all, in its condition of sociological theory, which is, by the way, ironic, since this same criticism has been directed by EM theorists against the concept of SD. The problem is perhaps not just in EM theory but rather in the very conception of the theory that is used to make this type of criticism. It is for this and other reasons that we will deal in the next chapter with the concept of SD. After all, no other concept in environmental sociology has been as controversial as it is. By analyzing this concept, not only will we have the opportunity to assess its relationship with EM theory, but we will also be able to gain a new understanding of EM as a concept.

It is also important to point out that the criticisms exposed above do not necessarily make EM theory unfeasible, but rather invite a reformulation of what EM theorists have written about it. Some of these points have already been discussed, and others we will try to clarify in subsequent parts of this book. EM theorists prefer to speak of "emancipation of ecology" and put aside a theoretical element that seems to be directly associated with environmental problems: capitalism. Thus, if economic growth must be reconciled with environmental protection, this occurs from a dimension of modernity that EM theorists dispense in their analyses. For all these reasons, EM theory is still permeated by a series of contradictions and ambivalences that can only be overcome once some of its assumptions are revised and reformulated. This, in our view, is not impossible. After all, as we have tried to show, a new theoretical direction for EM is both possible and desirable.

There are other very problematic aspects involving EM theory that were not mentioned in this part of the book but which will be addressed further with the concept of SD and the work of Giddens and Beck in subsequent chapters. Part of the importance of EM theory, and whether or not what they have to tell us makes the arguments that were woven in this chapter unfeasible, will be outlined concerning the next approaches that we will analyze.

NOTES

1. For a general assessment of the concept of EM, see also Andersen and Massa (2000).

2. This same argument can be found in Hajer (1995, 73).

3. Observing this question from another angle, Weale (1992) argues that the dominance of political elites in the formulation of environmental policy was gradually challenged as the environmental movement began to demonstrate greater technical

competence, which subsequently led to greater participation by environmental groups in the creation and formulation of environmental policy.

4. It is important to note here that the discourse of EM is seen as having its origins in the main document that will popularize the concept of SD within international environmental policy (Brundtland Report). What this interpretation suggests, then, is that SD and EM have a common origin. And that, for this very reason, they share many of the same assumptions that are present in contemporary environmental policy.

5. Other authors also refer to EM as an ideology rather than a discourse. See, for example, Rinkevicius (2000). Weale (1993) does not clarify whether or not there are differences between the various uses of the term. In fact, he seems to view them all as interchangeable. In any case, it is not uncommon in the social science literature to use these terms interchangeably (Purvis and Hunt 1993, 473). Thus, if we review the definitions of discourse (Hajer, Dryzek) and ideology (Weale), we will find that the differences between these definitions seem to be minimal.

6. "Story-lines" in the original. Dryzek (1997) himself borrows the concept of storyline from Hajer (1995).

7. The distinction made in this part between the old paradigm of environmental policy of the 1970s and the discourse of EM comes from the work of Boland (1994). In his text, Boland describes these differences in more detail and also provides a summary table showing these differences.

8. For an evaluation of EM as a modernizing theory, see Seippel (2000).

9. In chapter 4, we will explore Giddens's (1990) ideas on some of these issues and the relationship that can be established between his work and the theory (and discourse) of EM.

10. See, for example, Mol (1995), Spaargaren and Mol (1992) and Spaargaren (2000).

11. Or consider the words of Michael Jacobs (1991) in the following passage:

> Growth can take different forms. In some cases an increase in resource use and wasted emissions may be necessary, but in others, it may not. A firm which expands production of financial services, or recycled paper, or solar panels, may succeed just as well as one which builds nuclear power stations (probably better!) or imports tropical hardwood. Yet its impact on the environmnet may be much less. It may even improve it. It is *material* or physical growth which matters for the environment. But what the economic system requires is *financial* growth. Firms must expand their profits; they do no have to use up more resources. In theory at least, financial growth could still occur even if physical expansion were environmentally constrained. (26)

12. To see what greater efficiency might entail in each of these dimensions, see Jacobs (1991).

13. We will not make an extensive assessment of the specific characteristics of each of these instruments here. It is possible to find in Jacobs (1991) a description of the strengths and weaknesses of each one of them.

14. An assessment of this issue can be found in Mol (1995, 33), Spaargaren (2000, 50) and Leroy and Tatenhove (2000, 155).

15. See also the paper by Jänicke et al. (2000).

16. It should be emphasized that for Jänicke (1990, 95), both postindustrialism and superindustrialism are not "ideal" types of social development processes but represent the description of differentiated trends and possibilities that exist in industrial societies that are already taking place and that may be reinforced in the future.

17. Many EM theorists emphasize the need for technological change to reconcile economic growth and environmental protection. But as Gouldson and Murphy (1998) show, the response that companies can make includes not only a technological dimension but also organizational and strategic dimensions. Often, a change in one of these dimensions results in changes in subsequent dimensions. Due to space constraints, we will limit our discussion to only the technological dimension, which is highlighted in the literature from EM.

18. It should be noted that the relationship between these types of technologies should not be seen as antithetical, contrary to what some of the EM literature suggests. According to Gouldson and Murphy (1998), even if companies adopt preventive technologies on a large scale, it is unlikely that all pollutant emissions can be eliminated at their source. Thus, even if all emissions could be minimized, there would remain a role for a second phase of reactive technologies.

19. Literature on EM is still divided between those who trust the market and others who emphasize the role of the state. For this discussion, see Mol and Sonnenfeld (2000), Spaargaren (2000), Jänicke (1990), Leroy and Tatenhove (2000), and Gouldson and Murphy (1998).

20. Leroy and Tatenhove (2000) refer to them as rule-directed arrangements and rule-altering arrangements. For a debate on these different types of regulation, see also Mol, Lauber and Liefferink (2000). The theme of rule-altering arrangements, which has only recently been taken up by theorists of EM, also seems to be related to the theme of self-regulation and governmentality. Rule-altering arrangements are an attempt to give the state the power to influence economic activities without necessarily resorting to direct instruments. The issue of governmentality, analyzed by authors such as Neale (1997) analyze, is framed within this problematic. Governmentality, according to Neale, "denotes the emergence of a rationality of government which focusses not so much on the direct exercise of state power, but on a variety of processes by which the conduct and daily life of the population might be more closely regulated and monitored" (1997, 3).

21. I will not attempt in this work to examine how the EM seeks to balance the use of environmental policy instruments within its framework. However, it seems clear that the use of economic incentives gets a lot of space and projection in the EM as a political program. After all, it is the use of these instruments that is implicit in the process of "ecologizing the economy" and "economizing the ecology." Thus, if command and control instruments seem to be important for EM, it is obvious that this importance should be examined from the point of view of the process of "economizing the ecology," in which economic instruments tend to receive a larger projection. The reason for this is that technological and industrial change must work without betraying the beliefs and values associated with economic liberalism. This is a question to which I will return later.

22. According to Jacobs (1991, 152), it makes little sense to choose between one approach or another in an absolute way. Different instruments are appropriate for different circumstances and may eventually be used simultaneously.

23. For Jänicke's reflections on structural change in the economy and its benefits for environmental policy, see also Jänicke et al. (1989; 2000).

24. For comments on this study, see Simonis (1989), Gouldson and Murphy (1997; 1998) and Jänicke, Mönch, and Binder (2000).

25. We will not attempt here a systematic analysis of the differences between capitalism and industrialism, although this distinction is important for what we have to say below. Our aim in this work is only to try to point out the most obvious confusions EM theory creates with respect to these points. An author like Giddens (1985) makes the following distinction between capitalism and industrialism. For him, capitalism is defined as a system of commodity production centered on the relationship between private capital ownership and wage labor, in which production takes place in the context of competitive markets. Industrialism is associated with the use of inanimate material energy sources for the production of goods through the use of machinery and modern technology.

26. Recently, Spaargaren and Mol (2000) have presented a more careful view of capitalism. They do not claim that capitalism is essentially ecological, as neoliberals claim, nor that capitalism plays no role in environmental degradation, but rather that (a) capitalism is not static and can change to include environmental issues; (b) environmentally friendly production and consumption processes are possible with other "relations of production"; and (c) no alternative to capitalism itself has yet proven viable on economic, environmental, and social grounds.

27. See, for example, the work of Bell (1999).

28. For this reason, Gouldson and Murphy (1998) criticize the selective bias of EM theory, as it focuses only on the industrial, and not the capitalist, dimension of modernity.

29. For an assessment of theories of industrial and postindustrial society, see Badham (1986), Allen (1992), and Giddens (1982).

30. Although Giddens participated with Beck and Lash in a book entitled *Reflexive modernization* (Beck, Giddens, and Lash 1994), he makes it clear that he prefers to make use of the concept of institutional reflexivity rather than the concept of reflexive modernization. To Giddens (1994b), reflexive modernization, like the concept of modernization before it, contain within itself the assumption of a clear direction of development. Thus, Giddens rejects the attempt to see social change in modernity through a directionality that follows universal stages in the sense that they are processes that repeat identically in different parts of the world, whether in rich or poor countries.

31. See, for example, "Functionalism: après la lutte" (Giddens 1996). Apparently, EM theory carries a tension within itself that is expressed in its connection with modernizing and systemic theories and its approximation, especially in the work of Mol (1995), with Giddens's theory of structuration and modernity (1984; 1990).

32. Mol links the emergence of EM to the struggles established by environmentalism in the 1960s and 1970s. In *The Refinement of Production*, he notes that the

emancipation of ecological rationality in its ideological dimension began in the early 1970s. One of the factors that triggered this emancipation was the emergence of environmental discourse and ideology promoted by environmentalism (1995, 30). Thus, there is a parallel between the effects produced on the economic system by the labor and environmental movement. Just as the labor movement placed limits on the economic rationality of capitalism for social reasons, environmentalism would do the same with regard to ecological issues. The limitations imposed on capitalism for social reasons, as Mol (1995, 31) points out, occurred through struggles, conflicts, and disputes, and the same would have happened with the process that led to the emancipation of ecological rationality.

33. Parsons's view, in a way, expresses similarities to the evolutionist view of Herbert Spencer, which for some authors was a strong influence on his social theory. Spencer used the idea of social differentiation to explain the superiority of modern industrial societies over societies that preceded them in human history from an adaptive perspective. According to Crook, Pajulski, and Waters, "Spencer's notion of differentiation is a Darwinian evolutionary statement—a more differentiated society is a more advanced society, that is, a society which is better adapted to its environment because it has competed with other societies and outlived if not absorbed them" (1992, 3).

34. A caveat must be made to the arguments of Seippel (2000) and Buttel (2000b) since, by seeing the process of socioenvironmental change as a process of functional differentiation, EM theory could be associated with structural-functionalism (McLaughlin, 2021), and in that case, it would be possible to inscribe EM theory in a conceptual frame of reference that finds its place in contemporary social theory. However, it seems to us that it is not this characteristic of EM that makes it well received in the literature of environmental sociology. In fact, there are few works that make use of this structural-functionalist postulate to defend it.

35. For the condition of sociology as a discourse, see Alexander (1987) and Brown (1989).

36. Barry (1999b, 161) also emphasizes that the view of EM would be very close to that exposed by authors such as Jacobs (1991) and Eckersley (1995).

37. It should be noted that the approaches to regulation in the social sciences are numerous. Thus, knowing which approach to regulation is closest to ME's theoretical interests involves a more extensive analysis of these issues, which is not possible to do here. For a review of the literature on regulation theory, see Boyer (1995) and Jessop (1990).

Chapter 3

The Challenge of Sustainability

Sociology, Justice, and Democracy

In this chapter, we will work with the concept of sustainability. We will argue that sustainability is an eminently normative concept and we will try to evaluate from there its implications for the theory of EM and for environmental sociology itself. To do this, first of all, we will refer to the sociological criticisms addressed to the concept and then we will examine its normative and contestable character. It should be noted that, in this chapter, the terms sustainability and SD will be used interchangeably; however, efforts will be made to clarify the differences and similarities between the two.

THE CONTESTABLE NATURE OF SUSTAINABILITY

Weinberg, Pellow, and Schnaiberg (1996) note that, by virtue of the qualities that the social science literature attribute to the concept of SD, the latter started to be considered as uninteresting, if not useless, for sociological analysis. The authors point out that the following adjectives have been used to talk about SD: vague, empty, inaccurate, and expressionless. This view that the concept is "vague," "sweeping," and "meaningless" for the social sciences is not the result of the considerations of a particular author, nor does it express an exception in contemporary social sciences. This view, which seems to be predominant, can be confirmed in the work Richardson (1997), according to which SD is not just a political mistake; it represents a fraud as it attempts to obscure the contradiction between the finitude of the Earth and the expansionist character of industrial society. Lélé, when reviewing the literature on DS, concludes that the concept is in "real danger of becoming a cliché" (1991, 607). SD is a term to which everyone pays homage but nobody defines precisely. A central problem is the lack of semantic and conceptual clarity, which hampers serious debate about what SD really means.[1] As we will see, the reception of the concept of SD in many works of social sciences is characterized by an apparently critical and skeptical tone.

Jacobs (1999a) identifies at least three different responses given by social scientists to the concept of SD. The first is characterized by frustration and irritation and can hide a technocratic understanding of the concept by those who want to establish a unique SD definition. A second answer is the simple rejection of the term. A third answer comes from cultural critics who associate the concepts of modernism, technocratism, positivism, and simplistic scientific realism.. It is important to point out that many sociological works have questions about the concept rather than a complete rejection, a dubiousness that is usually not clearly expressed. If we look at some of the criticisms we saw earlier, we can see that however blunt and different they may seem, in many of them SD continues to be seen as something necessary and valuable to the social sciences. Lash, Szerszynski, and Wynne (1996), for example, despite their criticisms of the concept, point out that SD emerged to emphasize the importance of issues such as equity, justice, and human rights and also recognize that in its early stages it promoted a cultural and constructive relationship between the environment and society. They seem, well, to recognize positive points that can be associated with the concept. Although he criticizes the contr adictions of the concept, Lélé (1991) stresses that the SD has a certain "political force" (612). Likewise, Redclift (1987) does not fail to admit the importance of the concept. The latter argues that "the absence of any agreement about what 'sustainable development' actually means, still less whether it can be achieved in the real world, does not mean that the concept is useless, but it does mean that its use requires close attention. The idea of 'sustainable development' remains a powerful one" (Redclift 1992, 395).

It is curious to note that no matter how much criticism is addressed toward it, SD continues to be considered a "powerful idea" or as presenting a "political force." Few sociologists are predisposed to criticize it absolutely. The concept of SD presents problems, but apparently this does not mean we should discard it. So there is a tendency in the sociological literature to point out the contradictions of this concept but also to recognize its importance for one reason or another. Irwin (2001, 43) seems to be correct when he says it is possible to be ignorant of or ambivalent about the meaning of sustainability; on the other hand, he says, it is very difficult to stand entirely against it. Likewise, according to O'Riordan (1993), SD can show itself as a chimera, expressing all kinds of contradictions and be interpreted in the most different ways. However, as an ideal, SD is, he continues, "as persistent as a political concept, as are democracy, justice and freedom" (65). In fact, he adds, "it cannot be disconnected from these three ideals" (65). As we will see shortly, this vision has been increasingly recognized by many social scientists and has contributed to making sustainability a sociologically defensible concept.

Neither SD nor sustainability have yet received the attention they deserve in contemporary sociological literature. In certain works, what we can see is silence in relation to them. Other times, negative and skeptical criticism is the only noise. The political challenges of sustainability have been under-estimated at the very moment when it presents itself as a widely used term, carrying with it varied meanings. Fortunately, it is already possible to notice the change in attitude on the part of some important social scientists, who are beginning to recognize sustainability as an important topic of debate.[2] To better understand the problems and challenges the concept of sustain-ability poses to us, we have to review some of the misconceptions that may be behind the criticisms addressed toward it. Many of these are based on assumptions that must themselves be revised.

A frequent criticism, as we have seen above, addresses the diversity of definitions of sustainability. An answer to this is often the proposal to estab-lish a single and consensual definition. However, this way of looking at the problems surrounding the concept is not in itself consensual. After all, not all social scientists see this plurality as evil and not all of them link the importance of the concept to its analytical or descriptive capacity. For Jacobs (1999a), for example, the search for a single and precise meaning of SD is wrong. It is based on a distorted view of the nature and function of political concepts. The diversity of SD concepts should not be seen as a sign of lack of precision, but "such contestation *constitutes*," according to him, "the political struggle over the direction of social and political development" (1999a, 26). Similarly, Lafferty and Langhelle argue that the most significant potential of the concept is found not in the academic sphere but in its political dimension. The promulgation of the idea by politicians and bureaucrats is in inverse proportion to its rejection by critical social scientists, they note, adding that "denying the usefulness of 'sustainable development; as an analytic concept, or the attractiveness of it as a normative concept, does nothing to impinge on either its popularity or import as a political concept" (1999, 2)

SD is a contestable concept, along with many others from the social sci-ences, such as democracy, justice, freedom, power, responsibility, and inter-est, among others. What is common to most of these concepts is the fact that they show themselves as central to political life. Important concepts, especially those constitutive of modern political life, are often contested. The importance and strength of these concepts arise precisely from this contro-versy, something that is common, in fact, to the concepts of democracy and SD and other political concepts. At the same time, we should not overlook the fact that this debate occurs because of its connection with concepts such as democracy and justice, which are themselves debatable concepts. The debate that develops around concepts of this kind should therefore be regarded not as a "disease" but as a constitutive aspect of the political dispute that such concepts tend to generate.

Connolly (1983) conducted one of the best analyses of the contestable character of concepts of this type. As he notes, many social scientists work from assumptions that prevent us from recognizing that the controversy raised by such concepts is an inherent aspect of them. Among the assumptions are those that refer to the dichotomies between normative/descriptive, conceptual/empirical, and so on. Recognizing the essentially contestable character of these concepts help us understand them in terms that are more applicable to the political phenomenon, thus allowing us to grasp why central concepts of politics are so often the target of much conflict. Connolly establishes three conditions by which we can identify the emergence of a contestable concept: first, when it comes to a normative concept in which what is sought to be described is considered something socially valuable; second, when the practices that involve this measurable goal comprise an internally complex set of dimensions; and third, when the rules to operationalize such a concept are relatively open.

Essentially contestable concepts are those that involve "endless disputes about their proper uses on the part of their users" (Connolly 1983, 10). Thus, to say that a concept is essentially contestable is to say that the judging criteria it expresses are open to debate. This complexity, together with the moral dimension that pervades concepts of this type, makes it difficult to have a single, consensual view of them. Concepts such as democracy, justice, and sustainability cannot be evaluated by a single criterion, but by several. Furthermore, each criterion tends to present itself as multidimensional, generating an even more complex array of concepts. This not only increases the probability that authors may differ in their assessment of the concept, as some will focus on certain criteria to the detriment of others, but it also makes it difficult to fully operationalize the concepts as a whole. There is no specific solution to resolve the dispute that surrounds such concepts. For Connolly (1983, 40), reason can play an important role in this picture through scrutiny of particular concepts. Bryant (1995, 55), on the one hand, argues that in these cases, we should avoid conceptual variations that prove unnecessary; on the other hand, those that remain can be seen as a positive aspect of the debate, as it indicates that societies are open to new possibilities and opportunities for their constitution.

That said, our main interest concerning the concept of sustainability is to assess its normative dimension. It seems to us that a central dimension to understanding the contestable character of the concept and its importance as one for the social sciences is precisely to access the normative dimension it brings with it. As we have seen before, the moral dimension that pervades most contestable concepts is one of the reasons, along with conceptual complexity, that immerses then in a series of controversies. In *The Notion of Sustainability and Its Normative Implications*, Skirbekk shares this view. In the preface, Skirbekk explains that due to its complexity, sustainability requires an interdisciplinary vision. Skirbekk then argues that "the notion of sustainability is essentially normative, requiring an ongoing discussion about ethical priorities, both as to what we take to be a good life and as to what we regard as our obligation towards sentient beings and towards the biosphere in the long run" (1994b, 4). This view is also reaffirmed by authors such as Lafferty and Langhelle, who consider that the concepts of SD "are all aimed at the future; they are all normative in that they say something about how the future should be; and they admit 'permissible' development paths, depending on the scope of the definition" (1999, 25). As for Jacobs, "*no* concept of environmental protection is able to avoid value juggements" (1991, 78). To him, "sustainability is an *ethical* concept" (77). However, admitting that sustainability is a normative concept poses several challenges. First, if today there is growing agreement regarding the normative dimension of the concept, there are also different ways in which this dimension is understood. Rather than disagree on whether or not this is a normative concept, authors may differ on how the normative character of the concept should be assessed. This should come as no surprise, since morality is itself a contestable concept. What counts as moral, the objects of judgment, and the forms of justification are all presented in a pluralistic and diverse way (Lukes 1977, 173).

This normative aspect of the concept has been somewhat neglected by large part of the literature on sustainability. The authors mentioned above are some exceptions in the vast bibliography that exists on the subject. Economic approaches that seem to erase this dimension of the concept have predominated. A reduction of sustainability typologies into typically economic approaches, as McManus (1996) indicates, usually fails to consider the cultural bases for sustainability. Likewise, for Barry (1994b, 9), the politics of sustainability is antithetical to the current and scientific understanding of sustainability.

A preliminary assessment of the normative content of sustainability will be outlined below, without the pretension of providing a ready-made theoretical framework for these issues here. With the help of several authors, we will try to interpret the concept concerning its normative dimension. We will investigate whether the arguments related to the normative nature of sustainability are persuasive and their implications for EM theory and environmental sociology in general. We will try to point out the direction a normative theory of sustainability may take, without, however, closing the discussion in a conclusive and definitive manner.

THREE CONCEPTS OF SUSTAINABILITY

For Dobson (1998), there are two basic ways to deal with the concept of sustainability: One of them seeks to focus on the definition of the concept and the other is what he calls discursive, describing how the concept is used in political life. An example of this discursive strategy is found in the works that focus on the process of implementing SD.[3] These works suggest that given the numerous existing definitions of this concept, one option for analysis is to evaluate its uses in political life. From this perspective, sustainability is seen as a discourse permeating contemporary environmental policy, something similar to what happens with the analysis Hajer (1995) and Weale (1992) make of the EM concept presented in the previous chapter. However, considering sustainability (or SD) a discourse, as Dobson (1998) notes, seems to exempt social scientists from arriving at a precise definition of it. Thus, for example, Lafferty and Meadcrowcroft argue that "this study does not start from an autonomously derived (either logical or philosophical) interpretation of what SD 'really' means" (2000a, 17). A similar position is that of Baker and colleagues, who suggest that

> if attention is focused on sustainable development as a social and political concept, attention can be turned away from sterile debates about the precise meaning of the term, and focused instead on the contemporary process of implementing sustainable development policies and the alternative conceptions that are developing concerning how sustainable development should be interpreted in practice. . . . The focus is on how the meaning of sustainable development is interpreted in a variety of ways, developed into policies and programmes, and then reinterpreted in the light of the experience of implementation. . . . Sustainable development is thus mediated through the process of implementation. (1997, 7)

While the strategy that seeks a definition of the concept maintains the objective of defining what SD is, trying to fit it into a specific definition, the discursive strategy limits itself to describing how the concept is used by agencies, governments, and various social actors.

Problems can be pointed out in both strategies of approaching the concept. Seeking a precise definition for SD risks introducing a new conception among the large number of existing ones. The discursive strategy, in turn, does not point to any future direction of the concept, running the risk of simply reflecting its current use without highlighting its future weaknesses and potentialities. Due to these weaknesses, Dobson (1998) proposes another typological approach to sustainability, with the advantage of making explicit the components that each and every concept of it contains. This typology is structured from questions and answers that permeate the literature on sustainability. As Dobson observes, every conception of sustainability has an organizing principle, and this principle arises from the following question: What should be sustained? And, besides this question are others directly associated with it: Why and how should this "something" be sustained? For Dobson, all the concepts of sustainability that exist today can be summarized in just three types. Dobson's definitions for each of them are presented below.

(A) Critical Natural Capital

Critical natural capital is first and foremost a form of capital. Borrowing the notion from Marx, Dobson includes in the notion of capital all "raw materials, instruments of labour, and means of subsistence of all kinds, which are employed in producing new raw materials, new instruments of labor, and new means of subsistence (Marx, in Dobson 1998, 40). The natural dimension of this capital concerns the properties of the environment that are not produced by human beings. As the author observes, nature is "largely regarded as 'raw material' and thus as an 'economic asset' in Conception A of environmental sustainability" (Dobson 1998, 41). The "critical" dimension contained in this sustainability discourse is due to the preconditional character that these forms of capital have for human life and social practices. As Dobson clarifies,

> This conception of environmental sustainability (A) is concerned with sustaining a particular aspect or feature of natural capital, what is the best way of describing this aspect or feature? The answers to that question that emerge from the literature are undoubtedly best captured by the term "critical natural capital." . . . "Critical" here is to be understood primarily in terms of "critical to the production and reproduction of human life," and this points us in the direction of natural capital whose presence and integrity is preconditional for survival. (43)

So critical natural capital refers to environmental materials, processes, or services that are essential to human survival and well-being and that cannot be produced by human beings. This does not prevent them from being impacted by our practices or subject to our control.

(B) Irreversible Nature

In the irreversible nature discourse, what must be sustained are the processes or properties of the natural environment that are considered irreversible yet not necessarily vital to human survival and well-being, aspects of the environment that, once destroyed or consumed, can no longer be re-created at all. Irreversible nature, according to Dobson, concerns

> natural objects, naturally occurring substances, and organic and inorganic nature, whether individual or collective, and I shall specify these meanings as and when necessary. The idea that animates Conception B, simply, is that what should be sustained are aspects and features of non-human nature whose loss would be irreversible. (1998, 47)

By placing irreversibility as a pillar of this concept, this view of sustainability distances itself from those views that consider that natural capital can always, in principle, be replaced by alternative resources. The irreversibility thesis is thus opposed to the view that sees the substitutability as extendable to any form of sustainability.

(C) Natural Value

What is sought to be sustained in the natural value discourse are the particular historical forms of the environment. Dobson takes this definition from the work of Holland, who conceptualizes this type of sustainability as follows: "What is handed down and maintained does needs to retain in the process something of its original form and something of its identity: there need to be continuities of form, which constitute what may be called 'units of significance' for us, as well as continuities of 'matter'" (Holland, in Dobson 1998, 51).

In this discourse, what stands out is the historicity existing in aspects of the environment. Holland is defending, in the previous passage, the recognition that certain events and process of nature can be seen as particular historical phenomena and that they should therefore be valued as such. In short, "units of significance" for human groups are often associated with certain "continuities of matter." These two dimensions can, under certain circumstances, be linked with each other.

We will not be able to assess all three concepts of sustainability that we have seen above, nor will we be able to assess the normative dimension that underlies each of them. For reasons of space, we will focus on only one of the concepts of sustainability exposed by Dobson (1998). It is worth noting that we are not going to make a conceptual analysis here, but rather one that crosses the distinctions made by Dobson. This is possible since there is no barrier between the concepts of sustainability by Dobson and discursive approaches. To capture the normative implications of the concept of sustainability, we have to go beyond the typological approach provided by Dobson. As we aim to assess the normative content of sustainability, the best way is to also perform a discursive assessment. The reason we focus on the discourse is because, so far, there is no normative theory that allows us to relate these points systematically. On the other hand, some authors and works seek to provide a preliminary answer to the normative challenge of sustainability.

Let us quickly return to the issue of discourse we have seen in the first chapter. Dryzek (1997) defines discourse as a shared way of apprehending the world. Every discourse is embedded in some kind of language that enables those who subscribe to it to interpret information and organize it in such a way that it allows for the creation of coherent stories or narrative forms. Each discourse is based on assumptions, judgments, and statements that constitute the basic terms for analysis, debates, agreements, and disagreements, both in the environmental area and in others.

A discourse, then, presents assumptions and judgments that give it coherence and that differentiate discourse X from discourse Y. However, this coherence is often presented in an imprecise way, given the complexity of the issues that are associated with it. For example, regarding the concept of SD, Dryzek writes that it, "like democracy, is a discourse rather than a concept which that can or should be defined with any precision" (1997, 125). Therefore, discourse tends to be less precise than concept. It does not organize assumptions in a systematic and coherent way, as a scientific theory can do, although some authors point out that even the concepts are immersed in some kind of discourse. According to Alexander, discourse, not just explanation, is a major feature of the social sciences. By discourse, he refers to "modes of argument which are more consistently generalized and speculative than are normal scientific discussions." While explanation is more disciplined in the process,

> discourse, by contrast, is ratiocinative. It focuses on the process of reasoning rather than the results of immediate experience, and it becomes significant when there is no plain and evident truth. Discourse seeks persuasion through argument rather than prediction. Its persuasiveness is based on such qualities as logical coherence, expansiveness of scope, interpretive insight, value relevance, rhetorical force, beaty, and texture of argument. (1987, 22)

What we will do next, then, is to evaluate a discourse that presents a conception of sustainability that can be situated among the ones we have previously evaluated. The conception of sustainability we will focus on is conception A (critical natural capital), and the discourse that can be associated with this definition is that of the Brundtland Report. The importance of analyzing SD discourse within the Brundtland Report is twofold. First, it expresses a concept of sustainability that can be placed among the three sustainability concepts present in the literature. As we shall see, there is a clear parallel between the conception of the critical natural capital of sustainability outlined by Dobson (1998) and the SD discourse outlined in the report. In both, the importance of environmental aspects for basic human needs is highlighted. Second, the report also points to the interconnection between sustainability and its normative dimension, which has been very little commented on in the literature so far. Thus, the Brundtland Report continues to be one of the pioneering texts in highlighting the moral issues involved in the concept of sustainability. Not only does it seem to have been a pioneer, but for some authors, it is the main reference by which we can think about this issue.

The works of social scientists will also be mentioned, since, in our view, it would be a mistake to exclude their contributions to the discursive character that involves the concept of SD. Thus, sociology can be seen as contributing to social discourses with a specific discourse: the sociological one. Social scientists are not exempt from the discursive dimension that permeates modern societies, and thus participate with their intellectual activity in the consolidation of public discourses. Let us consider the relationship that Strydom offers to the role that the concepts of "rights" and "sustainable development" played in their respective times:

> Consider for example such concepts as "violence," "order," "sovereignty," "rights," and "state," or such concepts as "poverty," "economy," "growth," and "justice," or such ones as "ecological crisis," "risk," "sustainable development," and "responsibility." . . . In their respective times, these three sets of concepts proved to be politically, socially and culturally highly significant. Not only did they decisively stamp the socio-political semantics of their respective eras, the language and vocabulary that ordinary everyday people and politicians used to make sense of their world, but they also entered into a variety of more specialised semantic fields, including literature, the theatre, philosophy, and sociology itself. (2000, 18)

For Strydom, sociology draws its semantics from concepts such as these, which become the axis of public debate in their respective eras. Sociology does not stand apart from this process but is itself constituted by this discursive process. In this way, he says, what "sociology does . . . is to translate practical discourses in society and their semantics into something different, i.e., into sociology" (2000, 18). This means that authors who offer an interpretation of the discourse on SD, pointing out its problems and shortcomings, become part of the discursive field of the concept and contribute to its reconstruction. Although we will also refer to passages from the Brundtland Report, we will focus on the interpretation some authors have made of it, and at the end of the chapter we will attempt to assess its implications for both policy and environmental sociology. We will highlight two analyses of sustainability that we believe represent different ways of responding to the normative nature of the concept. Although both express a normative dimension, they maintain a link to conception A of sustainability (critical natural capital), described above. The differences between the two concepts and the problems that exist between them are discussed later in the chapter.

According to one of the approaches, the normative content of sustainability arises from its relationship with the theme of social justice. A quick analysis of the interfaces between social justice and sustainability will be enough to raise some important problems between the concepts of sustainability and the EM theory analyzed in the first chapter. This view can be delineated by the interpretation of the SD concept provided by Langhelle (1999), Haland (1999), and Lafferty and Langhelle (1999). The second approach is proposed by Barry (1994b; 1999b) and Jacobs (1997), who also view the concept of sustainability as normative but also associate it with the issue of democracy (although distributional issues are also recognized as implicated in the debate around the concept). Next we seek to assess these two ways of interpreting the normative dimension of the concept of sustainability.

We must warn the reader that the analysis we are going to make of the normative content of sustainability and its relationship with the theme of justice or that of democracy is quite provisional. On the other hand, we believe it will help to illuminate the normative character of sustainability and to extract some implications for the SD concept that we have analyzed previously.

SUSTAINABILITY AND ENVIRONMENTAL JUSTICE

As we have seen in the first chapter, the EM storyline lies in the possibility of ecological economic growth; that is, in the idea of making economics compatible with environmental protection. This is the major rupture between the EM discourse and previous environmental policy. Although the SD discourse seems to share this assumption of an ecological growth, several aspects differentiate it from EM. First, SD has achieved greater worldwide recognition than EM. Furthermore, for many authors, sustainability is a moral or normative concept as it unites moral issues with the theme of environmental protection. This is what is implicit in the storyline of the SD discourse. As Dryzek (1997) reminds us, the core storyline of the concept warns us that existing development aspirations in the world cannot be met by reproducing the pattern of development promoted by industrialized countries. This is because reproducing this development model would threaten the balance of the world's ecosystems. In the SD discourse, economic growth is still necessary, not only to meet the needs of the world's poorest but also because poverty is in many cases causally linked to environmental degradation. But if economic growth should be promoted, Dryzek notes, it should be guided in ways that are both environmentally benign and socially just, where "justice here refers not only to distribution within the present generation, but also across future generations" (1997, 129). In the context of this vision of development, it is clear that the environment is linked to concerns that have to do with human needs, especially the needs of the poorest, and that the process of meeting these needs, which can mean boosting economic growth, must be reconciled with ethical concerns related to the relationship between present and future generations.

Unlike the EM discourse, the SD storyline has strong normative content. To examine it, we will have to consider several aspects of the SD concept exposed by the Brundtland Report. SD is defined in this report as the "development that meets the needs of the present without compromising the ability of future generations to meet their own needs." (WCED 1987, 43). The report mentions two other key related concepts:

- the concept of "needs," in particular the essential needs of the world's poor, to which overriding priority should be given; and,
- the idea of limitations imposed by the state of technology and social organziation on the environment's ability to meet present and future needs. (43)

The SD concept, as expressed in the Brundtland Report, links development to an ethics of justice. The relationship between development and the environment is established through moral considerations involving the issue of human needs, giving the latter a distributive moral priority. This is a central aspect that we cannot lose sight of. While the concept appears to be simple, its implications can be profound in the way we define, in ethical terms, the concept of development. Development should be understood, according to the report, as a process of change that seeks to satisfy human needs, and this process should take place in global terms, with the needs of the poorest at the top of the development policy agenda. SD and justice, therefore, are part of the same process. As Langhelle writes, "Social justice can be seen as equivalent to the satisfaction of human needs, which in turn is what constitutes the primary goal of development in sustainable development" (1999, 140). So the concern of SD is not, at first, with the environment, but with basic human needs.

We will not address here the various existing theories of justice; suffice it to say that the concept of basic human needs is vital to many of them. As Kolm (2000) observes, the satisfaction of certain needs is necessary due to the very existence of the individual as such. These are fundamental physiological needs that, unsatisfied, compromise the proper functioning of the mind and body as well as the constitution of the human being as a person. Theories of justice defend the satisfaction of these needs from an existential justification similar to that presented for basic freedoms, which is also complementary to it. To Kolm (2000, 403) a basic freedom can be a basic need and both can be commonly complementary. The satisfaction of basic needs prevents basic freedoms from being purely formal.

Perhaps one factor that makes the concept of SD seem so indigestible to many social scientists is its insistence on the concept of basic human needs. Haland (1999) is one of the few who seek to discuss the meaning of human needs in the Brundtland Report and the dilemmas that surround it within the social sciences. While the concept of needs is central to the discussion of SD, the report does not spell out what is meant by basic needs and uses the term in different ways and at different levels, for example, the need for food, clothing, hygiene, healthcare, self-reliance, and cultural identity, among others. Furthermore, according to the report, "perceived needs are socially and culturally determined" (WCED 1987, 44).

Any theory of human needs faces several challenges. One basic challenge of a theory of this type is to be able to stipulate the existence of needs that are inherent to each and every human being and, at the same time, remain sensitive to the historical and cultural character of the realization of such universal needs (universalism dilemma/relativism). The Brundtland Report embodies the problem by arguing in favor of universal, present, and future human needs while stressing that they are historically and culturally achievable. Another challenge for such a theory comes from the division between basic and secondary needs. If needs are specific to time and culture, is it legitimate to specify a hierarchy for them?

For many authors, the questions posed to us by a theory of human needs are unsolvable. However, the idea of human needs plays an important role in contemporary political thought, and the possibility of building a theory of human needs has been revived in the last decade. One of the most important works in this regard is *A Theory of Human Need*, by Len Doyal and Ian Gough (1991), in which the authors observe that human needs are neither particular preferences that can be understood only by individuals nor static and understood only by planners or officials from different parties. Such needs are universal and likely to be known, but our knowledge about them, and about the necessary means to satisfy them, must be seen as something dynamic and susceptible to different influences and determinations (Doyal and Gough, 1991). In short, for Doyal and Gough, human needs guide what human beings must achieve if they wish to avoid threats to their physical and cultural lives. The authors envision the existence of two basic needs for every human being: the need for physical survival and the need for personal autonomy. In their own words:

> To be autonomous in this minimal sense is *to have the ability to make informed choices about what should be done and how to go about doing it.* This entails being able to formulate aims, and beliefs about how to achieve them, along with the ability to evaluate the success of these beliefs in the light of empirical evidence. . . . A person with impaired autonomy is thus someone who temporarily and seriously lacks the capacity for action through his agency being in some way constrained. (Doyal and Gough 1991, 53)

Elsewhere they add the argument that "since physical survival and personal autonomy are the preconditions for any individual action in any culture, they constitute the most basic human needs—those which must be satisfied to some degree before actors can effectively participate in their form of life to achieve any other valued goals" (Doyal and Gough 1991, 54).

What is important to retain from these passages is their implication for the concept of SD. This concept, as expressed in the Brundtland Report, although it does not present a complete theory of human needs, is supported by work in the social sciences such as that by Doyal and Gough (1991). The concept of SD in the Brundtland Report not only presupposes the existence of fundamental needs for human beings but also elevates these needs to the status of a normative principle for their vision of justice. The just society is one that, at a minimum level, satisfies the basic needs of those who are part of it.

Langhelle (1999) has made an interpretive analysis of the concept of SD in the Brundtland Report that helps us to clarify this question. In his interpretation of the report, Langhelle states that there is a close relationship between the satisfaction of needs and social justice. Social justice is seen as the process of satisfying human needs, and this process is understood as constituting the "primary objective of development in sustainable development" (1999, 140). Environmental sustainability emerges in this scenario as a material condition through which these goals can be achieved. In a sense, environmental sustainability allows the links between need satisfaction, development, and equity to be seen as integrated. Sustainability becomes necessary for people to meet their needs in the present, but it also becomes a condition for future generations to do likewise. The implication that the principle of sustainability presents for social justice involves two overlapping commitments to distributional issues. As the Brundtland Report notes, the concern the concept of sustainability raises for future generations is "a concern that must logically be extended to equity within each generation" (WCED 1987, 43). This shows us that social justice, understood as the satisfaction of needs, is at the heart of the discourse on SD. And this vision, which integrates sustainability and human needs, takes into account the interests of present and future generations. Thus, as Langhelle (1999, 140) points out, the concept of SD consists of two dimensions of justice.

This relationship between the satisfaction of needs and social justice raises two questions. The first is formulated by Ekeli (1999): Why does every individual have the right, as a matter of social justice, to have their needs met? The second is: What is the role of sustainability in this approach to justice as being equivalent to the satisfaction of human needs? Regarding the first question, the answer lies in the greater importance given by the SD discourse to human needs, to the detriment of human interests or desires. The normative force of the objective of satisfying human needs as a precondition for social justice precedes individual desires and interests or, in some way, constitutes the fertile soil for the latter to emerge. Thus, if social institutions are organized so as not to satisfy these needs, one can reasonably declare, as Ekeli notes, that these institutions are unfair.

But why, in this context, is environmental sustainability a requirement for social justice? The nexus between sustainability and justice is not merely contingent but engenders a theoretical and normative relationship in terms of principle. If we accept that human beings have basic needs that must be satisfied, that the satisfaction of these needs should be the object of the principle of justice, and even if, therefore, certain resources provided by nature are fundamental for this process to be conducted, then we will find that minimum sustainability is a preconditional requirement for the idea of social justice, as the satisfaction of human needs, to make sense. We can no longer believe that nature is a free and inexhaustible source of basic environmental services. The issue is not only, then, the fair allocation of resources (distribution), a common argument for theories of justice, but the maintenance of basic environmental services and resources for human well-being and survival. So, as Langhelle (2001, 16) indicates, the relationship between social justice and physical sustainability is not just empirical or functional, but theoretical and normative.

The concept of SD in the Brundtland Report presupposes a concept of minimal sustainability and the existence of aspects of the environment without which we cannot satisfy basic human needs. If we do not maintain a minimum quality standard for the atmosphere, soil, and water resources, the possibility of satisfying basic human needs could be compromised. So guaranteeing these environmental resources is not a purely economic issue, but a matter of social justice. Thus, if the economic system is an important area for the satisfaction of human needs and the conservation of these resources, it is not obvious that it is the *only* area in which these questions arise. These questions raise issues related to national and global governance that cannot be captured, for example, by an exclusive concern with the industrial dimension of modern societies.

The concept of minimal or physical sustainability presented by the Brundtland Report can be compared with what Doyal and Gough (1991, 157) call "satisfier characteristics." These are, according to the authors, elements that can contribute to the satisfaction of our basic needs in various cultural settings, although their form varies according to time and place. These satisfier characteristics refer to the properties of goods, services, activities, and relationships that enable human autonomy and physical health in all cultures. For this reason, they are also called intermediate needs, and among them the authors include economic security, physical security, and a nonhazardous physical environment (200).

The idea of physical or minimal sustainability, present in the Brundtland Report, is similar to the concept of critical natural capital formulated by Dobson (1998), cited above. This conception is interested in sustaining aspects of the natural environment that are considered "critical" to human survival. But here lies a problem inherent to the SD concept: If certain ecosystem services are essential to satisfy human needs and should therefore be considered essential goods to promote social justice, how can we define this minimal sustainability? Both the Brundtland Report and certain authors who defend the concept of SD expressed in it are evasive on this point, as its definition of physical sustainability is quite vague and general. The report states that, at "a minimum, sustainable development must not endanger the natural systems that support life on Earth: the atmosphere, the waters, the soils, and the living beings" (WCED 1987, 45).

Let us look at some indications of how we could conceive this minimal sustainability. In a negative view, minimal sustainability encompasses the prevention of any impact that could cause harm to human beings, both in the present and in the future. A minimal concept of sustainability implies preventing the emergence of environmental "threats" or "catastrophes" impacting human beings. Several threats could compromise the ability of future generations to satisfy their own needs. A weak or minimal version of sustainability requires, as Jacobs (1991, 71) indicates, that the environment is maintained to avoid an environmental catastrophe for future generations.

The Brundtland Report also presupposes a positive conception of minimal sustainability. At various times, it mentions the need to guarantee people and countries equal access to the planet's resources; that is, everyone should have guaranteed access to a minimum consumption standard for environmental goods and services. In this case, the minimal sustainability comprises considerations about distributive justice. The report provides the following interpretation of this relationship between minimal sustainability and distribution:

> Development involves a progressive transformation of economy and society. A development path that is sustainable in a physical sense could theoretically be pursued even in a rigid social and political setting. But physical sustainability cannot be secured unless development policies pay attention to such

considerations as changes in access to resources and in the distribution of costs and benefits. Even the narrow notion of physical sustainability implies a concern for social equity between generations, a concern that must logically be extended to within each generation. (WCED 1987, 43)

As we can see, sustainability involves issues of social justice both in the positive sense, regarding access to environmental resources, as well as in the negative, with regard to freedom from environmental threats that compromise the physical integrity and autonomy of human beings.

Thus, it is possible to define minimal sustainability by splitting it into two parts. First, it can be defined negatively; that is, as the effort to eliminate or reduce as much as possible the environmental risks to human life. Any human intervention in the environment is legitimate only if it does not pose risks to other people. However, such a view seems insufficient and can be complemented by including a more positive meaning related to the use of the environment. It could be defined as the ability to provide both present and future generations with equal opportunities for minimal environmental consumption (starting from basic human needs), without at the same time compromising the ability of the environment to perform its various functions.[4] Jacobs defines this more positive view of sustainability as the process by which the environment is protected in such a way that what he calls environmental capacities are maintained or preserved over time. These environmental capacities, which refer to the functions the environment performs for humans, must be maintained both at a minimum level to avoid future disasters and at a maximum level that allows future generations to enjoy the same level of environmental quality as current generations (Jacobs, 1991, 80).

Barry (1999b) refers to "sustainability" and "ecological rationality" indistinctly. He endorses the concept of ecological rationality provided by Dryzek (1987): the ability of ecosystems to consistently and effectively provide the best to support human life. What we want to retain from these two authors is their affinity with the concept of minimal sustainability present in the Brundtland Report and with the concept of sustainability as critical natural capital outlined by Dobson (1998), presented earlier. All of them refer to sustainability as a guarantee of providing basic environmental services to human life.

SUSTAINABILITY AND ENVIRONMENTAL DEMOCRACY

Another way to explore the normative content of sustainability is to examine its association with the concept of democracy. Just as a link can be established between sustainability and social justice, so can sustainability and democracy. Several factors bring sustainability and democracy together. First, both are contestable concepts. The existing conceptions of sustainability today are perhaps no less numerous than the ones we can find of democracy. Furthermore, both prove to be essential ideals for the current time. As O'Riordan (1993) notes, sustainability, as an ideal, is as important a political concept as are the concepts of democracy, freedom, and justice. Finally, there is a similarity in the popularity achieved by these two ideals in recent decades. In the same way that we live in a time when everyone calls themselves "democratic," we are at a time when everyone defines themselves as green, ecologists, and therefore defenders of some version of sustainability.

Sustainability and democracy have been seen as linked to each other. Munslow and Ekoko (1995) note that democracy is often identified as a condition for SD. This is because participation and empowerment of people would be constitutive elements of SD strategies, even though there are significant differences in the interpretations of this process. However, when reviewing the literature on the relationship between development and democracy, they find the existence of several theoretical models that establish the relationship between one and the other. The authors conclude that the existence of all these models shows that there is no predetermined correlation between democracy and SD. As in the view of Dobson (1998), it would not be possible, according to Munslow and Ekoko (1995), to establish a causal relationship between democracy and sustainability. The lack of a causal connection between one thing (sustainability) and another (democracy) indicates contingency here. Contingency may mean that democracy, as a decision-making process, does not provide guarantees for the realization of environmental sustainability. This view may in turn be expressed in the argument that sustainability should also be achieved through alternative means, means that appeal to "strong government" or even authoritarianism. I will address this last argument and return later to the first point related to the uncertainty of democracy in supporting sustainability.

The claim that democracy is incapable of achieving sustainability, and thus presenting the need to seek alternative means, implies in some sense that there is a possibility that sustainability can be achieved by nondemocratic means. To the extent the ecological crisis presents itself as serious and urgent, the demand for more participatory decision-making processes, with the attendant delays and complexity, would tend to compromise society's response to that crisis. There may be a certain technocratism behind this argument; if sustainability does not need to be built in a democratic process, then it must be left to technicians, specialists, and scientists. From this perspective, it would not be surprising to see democracy as a threat to sustainability, since "experts" have more knowledge to define and implement it. But there are reasons to believe this view is informed by misconceptions.

Decisions about sustainability are decisions of an eminently moral nature. Given the relative indeterminacy of the concept, deciding on a model of sustainability requires some sort of collective decision; that is, democracy. For authors like Barry (1999b), we can hardly assume a tenuous link between democracy and environmental politics, since democracy itself is a central and nonnegotiable value of ecological political theory. If sustainability is linked to critical natural capital, as we have already seen, what do we consider "critical"? If forests are essential to human survival, what kinds of forests will we preserve? How will the costs be distributed? Will we conserve only what is essential for human survival? How will decisions about distributional conflicts that exist in such decisions be encouraged? Can everything else that is not related to our most basic needs be left to destruction and excluded from a sustainability policy? There will likely be different answers to the question of how we want to conserve, and few would be satisfied with conserving only what is essential for human survival. Therefore, sustainability must be embraced by a policy perspective in which the various ethical priorities associated with it can inform the policy process.

Questions like these cannot be answered scientifically because their normative content requires intersubjective answers. This is not to say that scientific knowledge is not necessary, but it is not capable of providing ethical answers that require a political process. If we agree that promoting sustainability can lead to various moral disputes, there are good reasons, as Barry (1996, 119) argues, that the political process in question should be a democratic process. In this sense, sustainability is a flexible concept that presupposes the existence of norms and political structures that allow for some agreement on what should be promoted ecologically. For Barry, the concept of sustainability not only involves a debate about moral values but also suffers from essential indeterminateness. This requires that we understand concepts as discursively generated rather than as a given product. Thus he states,

Sustainability is thus both a matter of practical judgement, arising from its normative character, and a matter of knowledge. Sustainability is thus more than finding ecologically rational methods of production and consumption; it also involves collective judgement on those patterns. It is not just a matter of examining the ecological means to determined ends; ultimately sustainability requires a political-normative judgement on the ends themselves. Sustainability is therefore a matter for communicative as well as instrumental rationality, but the former takes precedence over the latter. This normative character of sustainability as a public principle or social goal makes it conducive to democratic as opposed to non-democratic forms of "will formation." (1996, 116)

Seeing sustainability through this prism raises a paradox. Since it is an imprecise social goal, sustainability consequently seems to suffer from a lack of
clarity. The concept tends to suffer, as Barry (1996) says, from an "essential
indeterminateness." Since the issues associated with sustainability raise a
number of moral dilemmas, it cannot be clearly and definitively defined from
this perspective. Although ecologists place a high value on sustainability,
it turns out that its value is indeterminate. This means that the ecological
dimension of sustainability must be calibrated in the implementation process,
taking into account participatory and distributive aspects related to the goals
it requires as a regulating social principle.

But what kind of democracy does the concept of sustainability require?
Barry (1996) sees the democracy as a process in which a collectivity dis-
cusses and decides, according to him, on principles and procedures designed
to govern its common life. It is also a type of political organization in which
the collectivity seeks a consensus on policies and forms of collective action
to achieve democratically chosen goals. The type of democracy closest to the
idea of sustainability would be what the author calls deliberative democracy.

Barry (1999b) seems influenced here by the conception of deliberative
(or discursive) democracy from Dryzek. For Dryzek (1990), deliberative
democracy is guided by communicative rationality, in which social interaction is free from domination, from strategic and manipulative behavior
performed by involved actors, and from (self-)deception. At the same time,
all actors must be equally capable of proposing and questioning arguments
(i.e., communicative competence). There should also be no restrictions on the
participation of competent actors. For Dryzek (1990, 15), under such conditions, authority tends to be based on the best argument that can be expressed
by the consistency of its description or explanation and understanding of the
empirical world, as well as by the validity of the normative judgments under
discussion.[5]

The choice of a discursive form of democracy is associated with the public
character of socioenvironmental issues. Sustainability can be seen as a public
good, the promotion of which raises distributional issues of various kinds.
Environmental services vital to human beings can be seen as public goods
in two senses. First, they must be collectively consumed and are therefore
indivisible. This means that the value of these assets cannot be assessed
individually. Furthermore, individual decisions could have consequences
for other parties involved. Second, these goods can be the subject of moral
debate, which makes it difficult to reduce their valuation to purely monetary
terms. Notwithstanding all these nuances and specificities involving public
goods, the forms and processes that exist today for environmental assessment follow a strictly economic profile: They include cost-benefit analysis
and other parameters that follow strict economic assumptions. The main

problem with these approaches is that they rest on the assumption that models used to evaluate private goods are also appropriate for evaluating public goods. However, such methods prevent or compromise the evaluations of the participants, since the latter are obliged to place their evaluation in terms of monetary cost-benefit, following an individualistic and private logic.[6]

Deliberative institutions, while not guaranteeing an assessment of environmental goods as public goods, make it more likely to happen. This is for three reasons. First, in such institutions, the arguments must be placed in terms of the general good. They are evaluated by taking into account the community or society in a more general sense and not the benefits associated with a particular group. Second, participants are forced to consider multiple points of view about an issue, which may cause them to revise their initial position. Third, deliberative institutions encourage recognition of the links between participants as well as greater solidarity between them. Experiences of this type have already been conducted in countries such as the United States, Spain, and Germany, and several contemporary studies show that in many cases, participants start to change their attitudes and preferences in the deliberative process. This suggests that such institutions do not seek to "reveal" the preferences of the people concerning the environment, which strictly economic approaches seek to do, but rather play a role in "constructing" such preferences.

Finally, it is noteworthy that deliberative democracy should be seen not as a replacement for liberal representative democracy, but as a complement to it. Deliberative democracy does not require participants to reach an absolute consensus, either; it only allows more general interests to be assessed and placed as a priority. Thus, we arrive at the hypothesis that if sustainability involves decisions about public goods, it requires institutions that allow citizens to make decisions together to obtain a collective judgment.

Having made these considerations, we can conclude that the concept of minimal sustainability presupposes three dimensions: a negative one, the absence of environmental hazards for human beings; a positive one, equitable access to a minimum of environmental resources and services; and a third one that incorporates the democratic-deliberative process for decision making. Although a conception of minimal sustainability presupposes the valuation of the environment in relation to the basic requirements of human life, not only may this aspect require public deliberation, but there may also be other forms of ecological rationality that should be considered (aesthetic, religious, etc.). The concept of minimal sustainability is, therefore, an anthropocentric concept, but it does not prevent nonanthropocentric interests from being considered. It is anthropocentric in formal terms, when placed as a process, but not necessarily in the content of the decisions that its structure promotes. Meeting this minimum sustainability standard is a requirement even for other forms of sustainability or ecological rationality to be satisfied. Although human beings are not the only moral subjects, they are the moral agents par excellence (Skirbekk 1994a). The decision as to what we are going to do with the environment is up to us, regardless of the moral status we may attribute to it.

BRINGING DEMOCRACY AND JUSTICE CLOSER

The view of sustainability as an imprecise normative concept that requires a kind of deliberative democracy seems to create some tensions with respect to the first view we have presented, in which sustainability is linked to issues of social justice. This is because deliberative democracy is strongly process-oriented and is the best means by which we can communicate our positions on the environment. However, it precludes any effort to establish substantive values for the environment in advance. Deliberative democracy is based on the idea of communicative rationality, and this places limits on those who wish to ascribe a priori substantive value to sustainability. In Dryzek's conception of discursive democracy, as Dobson (1996, 135) notes, communicative rationality is seen as a procedural standard that does not produce a final decision about the values to be pursued.

This is one of the reasons why there seems to be some tension both between environmentalism and democracy and between democracy and those who want to place a substantial value on sustainability. For writers like Goodin, "to advocate democracy is to advocate procedures, to advocate environmentalism is to advocate substantive outcomes: what guarantee can we have that the former procedures will yield the latter sorts of ourcomes?" (1992, 168). This pure and neutral proceduralism of deliberative democracy raises some discomfort among advocates of sustainability, as evidenced by the argument made by Jacobs. If we view democracy from a purely procedural standpoint, we are led to believe that a rational decision can be made on behalf of the common good. Jacobs then asks: "But surely this involves more than just a shared commitment to a procedure, to a communicative rationality? Does it not also require a sharing, at least at some level, of end-values? (1997, 227). Given this scenario, it is unlikely that a discursive democracy is sufficient to meet the conditions for sustainability. Even proponents of deliberative democracy, such as Dryzek (1990, 18), recognize that the procedural character inherent in the idea of communicative rationality that underlies deliberative democracy theory may prove problematic. As he notes, pure proceduralism becomes incoherent because engagement with the procedures of communicative rationality somehow implies acceptance of a way of life.

If deliberative democracy cannot be seen only in procedural terms, how can its connection with sustainability be established? The key lies in the question Jacobs asks us: How can an ethic of result, or the ultimate value of sustainability, be incorporated into the deliberative process? One answer Jacobs himself proposes is to place sustainability as a final negative value in the deliberative process, thus delimiting the possible decisions to be taken. This argument is also expressed in the work of Eckersley (1996), for whom the development of environmental rights could serve a similar purpose. Decisions on sustainability would thus be influenced by the consideration of environmental rights previously institutionalized in the deliberative process. In this sense, the dilemma between environmental values (sustainability) and democracy could be resolved not with more democracy, but rather with a new conceptualization of autonomy and justice and, consequently, with a reformulation of liberalism itself. Democracy and liberalism are connected in such a way that the former would have no foundation were it not for the liberal principles of autonomy and justice. In the words of Eckersley,

> If we are to give moral priority to the autonomy and integrity of members of both the human and non-human community, then we must accord the same moral priority to the material conditions (including bodily and ecological conditions) that enable that autonomy to be exercised. By widening the circle of moral considerability, humans, both individually and collectively, have a moral responsibility to live their lives in ways that permit the flourishing and well-being of both human and non-human life. This more inclusive notion of autonomy would necessarily involve the "reading down" or realignment of a range of "liberal freedoms" in ways that are consistent with ecological sustainability and the maintenance of bio-diversity. (1996, 223)

From this point of view, the disagreement between environmentalists (concerned with the ends) and democrats (concerned with the means) would have its origin not in democracy itself, but in the meaning of the concepts of autonomy and justice. Thus, a reformulation of the democratic project to make the goal of sustainability feasible requires a concomitant review of liberalism itself. Although he does not present an ecological conceptual reconstruction of these concepts, Eckersley (1996) shows how this might be done. If we give moral priority to autonomy, to the integrity of members of the human and nonhuman community, then we should give equal priority to the material conditions that enable the exercise of that autonomy. Recognition of the material basis of human autonomy could be justified as follows. One approach would be to argue, as Eckersley does, that certain basic ecological conditions are nonnegotiable with majorities because these same ecological conditions are ecological preconditions for the democratic participation of present and future generations. These ecological conditions that support life "might be seen as even more fundamental than the human political rights that form the ground rules of democracy" (1996, 224). This is not to say that certain ecological conditions are more important than political rights, but that such conditions are the basis or prerequisite for rights to be exercised. For if political rights are important to democracies, one must examine the social and ecological conditions under which they can be exercised. Thus, one aspect cannot be divorced from the other. Environmental rights and political rights are inextricably intertwined and should therefore be recognized together.

If we conclude that certain ecological conditions are fundamental to human beings, then we should try to translate them into environmental rights. For example, people might have the right to an environment free from threats to their physical integrity and autonomy, so that ecological security is as important as social security. This view of Eckersley (1996) is very close to that of the Brundtland Report (WCED, 1987). After all, the concept of SD, as expressed in this report, also links sustainability and basic needs such as physical integrity and autonomy.

As we can see, approaches that link sustainability to justice and democracy are not necessarily opposed, nor do they need to conflict with each other. Eckersley (1996) demonstrates that they are dependent on each other. Both the Brundtland Report and Eckersley recognize that the resources that the environment provides us are essential to meeting basic human needs. The author even acknowledges that her view would be implicit in the Brundtland Report. Some international documents and works support the idea of guaranteeing environmental rights to human beings, and an example of this, for her, is the Brundtland Report. The report also recommends a set of legal principles for SD, such as the one stating that "all human beings have the fundamental right to an environment adequate for their health and well-being" (WCED 1987, 348).

The idea that ecological values can be an extension of human rights, as Hayward (2001, 119) observes, is developed in the Brundtland Report, which presents the goals of environmental protection as an extension of the discourse on preexisting human rights. But what relationship can we establish between sustainability and human rights? What does this mean for environmental policy and for environmental sociology itself? These questions are complex and require separate studies. However, we will give a brief indication of the political and sociological implications. The recognition that certain environmental processes are essential to the satisfaction of basic human needs has major implications for environmental policy and also for sociology itself, because it leads to the realization that society's major institutions should internalize this principle. As Hayward indicates,

What it means to incorporate ecological values into political theory at the level of basic normative principles, I want to suggest, is, firstly, to treat environmental services and resources as social goods whose distribution is a question of justice, on the grounds that they represent generalizable interests warranting recognition at the level of basic institutions. It also means entrenching a recognition that not all "environmental goods" can or should be treated as resources or services, and that there should therefore be substantive restrictions on the utilization of certain environmental goods (2001, 118).

If such principles deserve to be included in the basic institutions of society, they should be addressed at the constitutional level. At this point, the relationship between sustainability and human rights is established. In establishing final values in the decision-making process, some decisions should be excluded because they violate the principle of sustainability if the idea of environmental rights is used to guide environmental policy decisions. In short, a minimal concept of sustainability associated with the minimum conditions for the realization of people's autonomy should be institutionalized and not left to the deliberative process itself. In this context, sustainability could function like human rights in democratic debate: "as a 'trump' which overrides contrary outcomes" (Jacobs 1997, 227).

Some authors believe that ecological values such as sustainability could be fully accommodated within existing human rights. After all, if sustainability implies some kind of "ecological security" or a "nondegraded environment" for human beings, these goals harmonize well with the right to health, life, and so on. On the other hand, authors such as Eckersley (1996) speak of the possibility of a fourth generation of rights, yet to be created, which she calls environmental rights. Regardless of whether the concept of sustainability can be accommodated within existing human rights or whether it involves the creation of new environmental rights, the moral basis of the concept may be justified by this kind of political and moral discourse.

Hayward's arguments are important once again in considering this issue of sustainability. According to him, social and economic rights can help protect environmental standards through human welfare. Thus, in certain circumstances, many issues related to health and decent living and working conditions can be affected by environmental conditions. Rights violations can be viewed as circumstances in which the conditions for decent living underpinned by human rights discourse are threatened. For Hayward, "The right to life might be deemed, more generally, to include the right to live in a healthy environment, a pollution-free environment and even an environment in which ecological balance is protected by the state" (2001, 199). In other words, preexisting civil and political rights can be mobilized to promote change toward an environmentally sustainable social order because they support environmental groups that oppose environmental degradation. Thus, when environmental degradation threatens basic human needs underpinned by human rights discourse, it is seen as a threat to those same rights. In this sense, there may be a strong affinity between the environmental movement and other social movements that advocate for democracy and justice. And if the concept of minimal sustainability is intertwined with that of social justice, one can conclude that the discourse of rights may be the best way to translate such environmental injustices into sociological terms. As Cooper observes, "it is common practice to describe violations of rights as acts of injustice" (1995, 141). Environmental sociologists who attempt to describe violations of environmental rights are thus helping to uncover and communicate acts of environmental injustice to the public, and these situations can be studied as violations of the ideal of sustainability. This could indicate that ethical issues, especially those associated with the distributional conflicts associated with environmental risks, are being separated from the industrial efficiency agenda. This is because ethical issues are generally perceived as an obstacle to instrumental rationality. For questions of this kind involve a communicative rationality and not an instrumental rationality.

Recently, certain authors have pointed out the need to reconnect sociological work with the theme of social justice and have also pointed out the need to base sociology on a theory of human rights.[7] Feagin and Vera (2001, 252) state, for example, that the moral imperative of sociologists must go beyond the borders of the nation-state to be guided by the tradition of human rights on an international scale. Investigating the excesses of power of nation-states and large corporations requires, according to them, an audacious moral position that endorses the human rights of all citizens on the planet. Turner (1993) also argues that sociology should comprise another one, of human rights, that transcends the normative sociology of citizenship conventionally embraced by social scientists.

In short, sustainability is at the same time a question of justice, democracy, and human rights. We could say, then, that an environmental sociology founded on the concept of sustainability is a normative sociology that has, in human rights and the very idea of (deliberative) democracy, a moral basis that guides its research and its practical guidelines for environmental public policy.

ECOLOGICAL MODERNIZATION AND SUSTAINABILITY

EM theorists subscribe to many of the criticisms that are addressed toward the concept of SD. Spaargaren and Mol (1992, 333) argue that the current consensus around the concept of SD is only possible because "sustainable development is a rather vague concept that allows many interpretations."

For Spaargaren and Mol (334), the concept of EM examines the relationship between the industrial system and the environment in the context of the most industrialized societies in the world, while the concept of SD projects its concerns also to the poorest regions and countries as well. Thus, there is a difference in geographic and analytical scope here. At the same time, the concept of SD, unlike EM, would seek to do so, they claim, while incorporating concerns related to the issue of equality and peace on the planet. This obviously indicates that these issues are not seen as related to the concept of EM. Despite these differences, they claim that both concepts have the same view of the environment and modernity. In addition to these arguments, Spaargaren and Mol support the view of Simonis, for whom the relationship between society and the environment under the conditions of modernity calls for *"industrial restructuring of sustainable development, or ecological modernization"* (1989, 361).

In this view, EM is seen as the SD industrial restructuring process. Obviously, these arguments serve not only to delineate the differences between EM and SD, but also to free the former concept from eminent ethical and political issues that intrude into the latter. The concept thus remains linked to an agenda of industrial system efficiency, where ethical issues related to development are outsourced to the concept SD. This could be used to justify the notion that the agenda of EM is more limited and focuses on issues restricted to the industrial system, making it easier to understand and apply..

For other authors, these discourses should not be mixed or integrated. Langhelle (2000) and Lafferty and Meadcrowft (2000b), for example, recognize that EM and SD are very similar in many aspects.[8] However, they state that there are substantial differences between them, with significant impacts on the policies inspired by each of them. The set of differences that these authors raise concerning these concepts is summarized in table 3.1.

By making these distinctions between EM and SD, the authors simultaneously criticize the properties associated with EM, probably because they believe the SD discourse is more sensitive to important issues related to the development agenda. But it is worth problematizing these distinctions between SD and EM. First, some of them are used by theorists of EM to defend it. According to them, features such as national scope do not diminish the importance of EM, but, on the contrary, give it more consistency. Another aspect that should be highlighted is that the criticism Lafertty and Meadcrowft (2000b) make of EM is supported by authors working with the concept of EM. Unlike the former, however, the latter consider a reformulation of EM to be possible. Hajer (1995), for example, a critic of certain technocratic versions of EM, proposes an EM model that he calls reflexive or strong EM. Christoff (2010), in turn, speaks of a "strong" EM, which would

Table 3.1. Differences between sustainable development and ecological modernization

Sustainable Development (SD)	Ecological Modernization (EM)
Political/normative concept	Analytical concept
• Global Scope: concern with global environmental problems and global ecological interdependence.	• National scope: concern with circumscribed environmental problems (regional and national level).
Concern for environmental and social justice. Interconnection between sustainability and justice issues.	Concern with efficiency
Third World: concern for poor countries.	West: concern for rich countries
• Requires structural economic change (economic growth is subject to SD).	• It does not face systemic aspects of capitalism.
• Emphasizes the role of government.	• Admits an unregulated economy.

Source: Elaborated by the author.

be institutional/systemic, communicative, democratic-deliberative, and international. In other words, criticism of the EM discourse does not necessarily lead to its rejection or to the impossibility of imagining alternative models. In fact, some alternative EM models might even be compatible with the SD concept. Third, some of the distinctions in the table may be controversial, for example, the claim that EM has a strictly national focus might be questioned. For Weale, "the new pollution policy," which for him takes shape from the precepts that form the discourse on EM, "now has an inescapable international dimension" (1992, 187). According to him, "the internationalization of pollution issues during the 1980s also reflects the growing appreciation of the fact that pollution itself is often an international problem" (187). In short, for Weale, EM emerged from the recognition of the global scale of various environmental problems and, in a sense, represents a reconfiguration of environmental discourse based on this recognition.

There are other questionable distinctions, such as the one that considers SD a more political concept than EM. As Eckersley observes, "shifting the discourse from 'sustainable development' to 'ecological modernisation' does not alter the fundamentally normative character of the environmental debate" (2000, 233). Therefore, EM can hide its moral implications, but that does not necessarily make it a less normative concept than SD. In this case, SD would be a more transparent concept with the normative implications that it, as a guide for environmental policy decision, offers. Furthermore, Spaargaren and Mol's argument that "sustainable development is a rather vague concept that allows many interpretations" (1992, 333) is not credible either. According to Christoff, the use of the EM concept also "varies considerably in scope and meaning" (2010, 212). Thus, if today there is a conceptual pluralism around sustainability, this pluralism also permeates the concept of EM. As a result, some authors argue that EM may suffer from the same conceptual impreci-sion as SD (Buttel 2000b, 61). From an ideological point of view, these con-cepts have interfaces: concern with ecologizing economic growth, promotion of green technologies, and association with governmental and international organizations. That is why Barry (1999b, 139) states that EM, at the regional and national level, is similar, in both origin and function, to SD at the inter-national level. These concepts represent similar responses for different levels of government (state and supra-state) and both have links, therefore, with the discourses of bureaucratic administration.

The important thing to remember is that both SD and EM, although distinct types of ecological discourse, maintain political affinities. The main differ-ence between them resides, at least initially, in the geographic and institu-tional scope of each. While the EM discourse turns to the regional/national level, the SD discourse reaches the international level. The former deals with the limitations and possibilities of action of the national state regarding environmental problems, while the latter starts with these limitations aimed at the supra-or transnational dimension. The most fertile analysis consists not in contrasting the local with the global but in analyzing the possibility of integrating them. And apparently, EM and SD seem to outline such a pos-sibility. The success of the ecological policy from EM is closely linked to the effectiveness and impact of the SD discourse at the global and transnational levels, and vice versa. As Young (2000) points out, the future perspectives for EM are also linked to its relationship with SD. At the same time, the ethi-cal agenda associated with the concept of sustainability need not exclude an environmental efficiency agenda associated with the need for transformation of the industrial system. They may conflict when the second agenda tends to cloud the issues of the first. Or when the latter (environmental ethics) tends to underestimate the need for the former (environmental efficiency).

After highlighting some similarities between SD and EM, let us point out their most fundamental differences. The opposition is not primarily between them but in the notion of sustainability that underlies each of them. This is not to say that the differentiation between these concepts is reduced to this topic, but it is an important starting point.

Every theory or discourse of sustainability must answer questions related to "what," "how," and "why" something in the environment must be sustained, or at least offer a process by which questions like these can be answered. Since such questions encompass moral and political content, we should rec-ognize that every concept or theory of sustainability expresses a normative political theory of some kind. So far, we have evaluated two approaches that seem to provide us with some answers to these questions, but these are only very preliminary indications.

We have seen that a theory of deliberative democracy is not sufficient to respond to the challenges of the concept of sustainability. Barry (1996), who sees deliberative democracy as a way to resolve conflicts in decisions related to sustainability, recognizes that democracy and justice should be integrated into any concept and political project of sustainability. His view is very close to the thinking of Eckersley (1996), which we have already seen. According to Barry, when it comes to the outcomes of political decisions, we cannot consider only democratic criteria, because these criteria refer to procedures and not to substantive outcomes. Therefore, he notes, we must turn to the justice to evaluate those very results (1996, 127).

Therefore, a normative theory of sustainability must integrate both a theory of justice and a theory of democracy, a fact that the SD discourse recognizes, while the EM discourse and sociology seem not to pay much attention. It remains for us to discover what, in both the EM discourse and sociology, opposes to this view. The first point to be unveiled concerns the concept of sustainability underlying the sociological theory of EM. Ecological modernizers use as synonyms for sustainability concepts such as support base and ecological rationality. EM is sometimes defined as the process of emancipation from the ecological rationality. However, Mol (1995) does not clarify what he understands by ecological rationality, under the justification that such concept has already been used by several authors. Among them, Mol gives us the example of Dryzek (1987), which seems to be his source of inspiration, or at least one of the sources for understanding the ecological rationality implicit in EM theory. So let us examine the concept of ecological rationality contained in Dryzek's work in order to return to EM below.

According to Dryzek (1987), natural systems can be valued in several ways: productive, aesthetic, religious, and scientific terms. Any form of functional rationality implies some value (or several) and an appropriate mode of behavior (or several) to achieve that value. Social structures express functional rationality insofar as they turn to certain ends. Thus, functional rationality constitutes an assessment standard embedded in social systems. As Dryzek (25) indicates, a rational firm produces profits, an economic system satisfies consumer demands, a rational-legal system resolves disputes, and so on. Likewise, he proposes a minimal conception of ecological rationality. In this minimal definition, ecological rationality is based on the "productive, protective, and waste-assimilative value of ecosystems—that is, those aspects which provide *the basic requirements for human life*" (34). Dryzek, then, defines ecological rationality as "the capability of ecosystems consistently and effectively to provide the good of *human* life support" (36). "Consistency," as Dryzek informs us, is about long-term sustainability: the ability to last and remain in its original form over time. In other words, the well-being of present generations should take no precedence over that of future generations when we recognize that certain environmental goods are fundamental to human well-being. Dryzek calls his conception minimal ecological rationality, as it focuses on valuing the environment in guaranteeing the basic prerequisites of human life. This does not mean that other ecological rationalities cannot be established in addition to these basic requirements, but they can be delineated only once the minimum ecological rationality is established. This can be seen as a basic rationality, without which other forms of rationalities cannot thrive.

Some criticize the concept of SD precisely for associating sustainability with basic human needs, seeing an exaggeratedly anthropocentric bias in it. However, even authors who defend a sociology of the emancipation of ecological rationality (EM) fall into this same point. If it seems inappropriate for the social sciences to rely on the notion of human needs (SD), what advantage would there be in supporting the idea of ecological rationality based on the "basic requirements for human life" (Dryzek 1987, 34)? Dryzek's concept of "ecological rationality," and of EM theorists, operates the same appeal to human needs as the SD concept seems to do. After all, what would be the difference between "basic human needs" (SD) and "basic requirements of human life" (EM)? Indeed, developing a theory of human needs is not impossible and, in some ways, is implied in the human rights discourse endorsed by most nation-states today. The validity of human rights is the precondition for the vital basic needs of individuals to be institutionally protected against specific threats caused by the state and the global order (Follesdal, 1999).

The real problem in the "ecological rationality" implicit in EM theory is to dissociate the concept from issues such as justice and democracy. Therefore, Barry's (1999b) critique of the ecological rationality present in Dryzek (1987) is pertinent, and we can extend it to EM theory. According to Barry, the conception of ecological rationality by Dryzek is problematic from an ecological point of view, as it refers only to the "human life suport with no reference to other values such as democracy, autonomy or social justice" (Barry 1999b, 108). This makes the criticism that has been directed at EM understandable, since its concept of ecological rationality seems to incorporate the same limitations that Barry (199b) associates with Dryzek's (1987) concept of ecological rationality. The neglect of important normative issues, associated with democracy and justice and directly linked to the sustainability policy, seems to be a problem related to the way in which ecological rationality in EM theory has been conceptualized.

CONCLUSION

Throughout the analysis of the concept of sustainability that we have performed in this chapter, the discourse and sociology of EM has presented two shortcomings: dissociating the issue of sustainability from the theme of social justice and sustainability from democratic institutional requirements. EM has been silent on the type of democracy needed to promote the goal of sustainability and has restricted itself to the debate on state regulation. While EM relies on fundamentally economic institutions, SD relies on political institutions such as democracy and human rights. Although EM has a more strictly economic bias, it does not necessarily slide into narrow economism, although this may be true for some of its versions. Dryzek (1997, 143) seems to be right in arguing that EM has a more specific focus on what can be done with the capitalist political economy within the limits of the nation-state. On the other hand, normative issues are much more prominent in the concept of sustainability or tend to appear more evidently in the debate about this concept, which does not mean that they are less important for EM theory and discourse. In any case, if the EM literature brings out more clearly the possibilities of ecologizing economic growth, there seems to be no reason to think that this goes against the defense of human rights and the establishment of an ecological (deliberative) democracy. It is worth remembering that in 1986, the UN established a third generation of rights aimed at developing countries, focusing on the "rights to development" (Eckersley 1996, 220).

The controversy surrounding the concept of sustainability is nothing new in the social sciences. In a way, theoretical and conceptual diversity is an endemic aspect of sociology itself. Practical difficulties of sociological research are generally attributed to the various sociological languages employed in different theoretical approaches (Stehr, 1982). A recurrent trend in sociology has been to see this theoretical diversity as something pernicious to the social sciences. Criticism of the imprecise character of the concept of sustainability (or SD) thus seems analogous to the critique conventionally made of the "pre-paradigmatic" and "immature" character of sociology. But, as we have seen, criticizing the concept of sustainability in this light reveals a misunderstanding of the problems inherent in the concepts coming from sociology and the social sciences. The concept of sustainability is essentially contestable, and EM and SD represent possible interpretations of it. Thus, EM and SD are part of the conceptual pluralism that marks the concept of sustainability in general. Furthermore, the multiplicity of the concept of sustainability arises precisely from its normative character; as it is a socially valuable principle, there is no prior and absolute consensus on its realization.

In this chapter, we looked only at the conception of sustainability as critical natural capital. As we have seen, even though there are different interpretations of sustainability, there is a growing consensus on the need to stipulate a minimum concept for it. Minimal sustainability is what is presupposed in the EM idea of ecological rationality, exemplified by Jacobs's "green economy" (1991). It is also implied in the SD discourse in the Brundtland Report (physical or minimal sustainability) and is defended by Barry (1999b) and Dryzek (1987). All these derivations emanate from a more general matrix, coined by Dobson (1998): sustainability as maintenance of the critical natural capital. This is supported by other works. By evaluating different approaches such as free-market environmentalism, EM, the ecological economics proposed by Jacobs (1991), and the approach of constant natural capital by David Pearce, Labaras (2001, 81) finds that a minimal common conception of sustainability can be found.

One important conclusion we can draw from our analysis of sustainability is that this concept seems to require a commitment to both a kind of democracy and the discourse of human rights. While the EM theory associates environmental sociology with an ecological restructuring of industrialism, the concept of sustainability in SD, as exposed by the Brundtland Report and by other authors, suggests that an environmental sociology would not be possible without adherence to a theory of environmental human rights. In this sense, environmental sociology is eminently normative, looking for its source of inspiration in deliberative democracy and basic human rights. This argument, although it may sound strange to conventional debates about sustainability and environmental sociology, proves to be coherent and persuasive when we consider the normative implications of the very idea of sustainability.

NOTES

1. This is a type of critique that EM theorists also make of the concept of SD. See, for example, Spaargaren and Mol (1992, 334).

2. For an examination of this argument, see, for example, Barry and Wissenburg (2001, 2) and Doherty and Geus (1996, 13).

3. This type of analysis can be found in Baker et al. (1997) and Lafferty and Meadowcroft (2000a).

4. This definition is slightly different from that of Jacobs (1991) because it considers an equal opportunity for present and future generations in accessing a minimum consumption standard necessary to satisfy basic human needs. This minimum consumption should be regulated taking basic human needs as a parameter and the capacity of environmental services themselves to perform their functions. Both are parameters because the satisfaction of needs cannot be accomplished without

considering the maintenance of environmental capacities and vice versa (Jacobs, 1991, 99–100).

5. For an analysis of the relationship between deliberative democracy and environmental issues, see Smith (2003).

6. To learn more about the predominance of economic approaches in environmental assessment, see Grove-White (1997).

7. The need to link sociology to a theory of human rights is explored by Feagin (2001), Feagin and Vera (2001), Turner (1993) and Sjoberg et al. (1995).

8. Some authors even see the EM concept as a substitute concept for SD. See, for example, Boland (1994).

Chapter 4

Ecological Modernity

Risk, Science, and Politics

In this chapter we will discuss the social theory of Ulrich Beck and Anthony Giddens. We will see that the work of these authors raises important questions for EM theory and the concept of sustainability that we have analyzed in the previous chapters. Initially, we will quickly review the role played by science and technology in the literature on sustainability and EM. We will then review the work of Beck and Giddens concerning this topic and its more general implications for the concepts of SD, EM, and environmental sociology.

SUSTAINABILITY AND THE SCIENTIFIC PARADOX

The problem in the concept of SD, according to some authors, is not in its political-normative nature, but in its pretension of positioning itself as a neutralist scientific discourse.[1] For Moser (1995), the concept of SD, as exposed by the Brundtland Report (1987) precisely demands that science and technology play a fundamental role in the political decision-making processes. In the SD view, science and technology are presented as clearly neutral means to achieve irrefutable political goals, such as growth, progress, and development. Despite a growing public recognition that science and technology are associated with many environmental disasters, Moser denounces that the DS discourse tends to avoid this problematization. Sachs aims criticism in the same direction, arguing that the attempt to reconcile environment and development, in the concept of SD, would have implied the emergence of the view that the world can be saved "by more and better administration" (1993, 11).

Dependence on science and technology is also associated with the EM discourse. Certain authors recognize that EM, both in theory and as a political program, is a tributary of scientific knowledge. For Cohen, EM, as a political program, it is "dependent upon a firm commitment to science and a preference to address environmental problems in technological terms" (2000a, 77). In other words, it demands great confidence in science and technology as a means of promoting the goal of sustainability. It is for this reason that some of its critics tend to regard it as a technocratic approach to environmental problems.[2]

The dependence of both SD and EM discourses on science and technology apparently places them in an opposite position to the theory of RS defended by Beck and Giddens. This tension has been highlighted by the theorists of EM itself. Mol and Spaargaren (1993), for example, claim that the theory of EM can be seen as opposed to the RS thesis because, unlike the latter, it offers a constructive approach to dealing with the environmental crisis. It would do so by assigning, unlike the theory of RS, a central role to modern science and technology in addressing the environmental crisis. RS tends to contradict EM theory because of its pessimistic view of the role of science and technology (Mol and Spaargaren 1993, 433). This difference between these theories is so perceived because, as we have seen, in EM theory the transition to a more sustainable industrial system tends to require a process of innovation in which science and technology play a central role. RS theory, on the other hand, takes a more ambivalent view of the role of science and technology in environmental policy.

Cohen (1997), following Mol and Spaargaren (1993), suggests that EM and RS can be viewed as different paths of change that developed countries can take. Cohen suggests that "the direction in which a particular society will progress [EM or RS] will depend on its predisposition to scientific rationality" (105). RS and EM theory would thus represent different models of social change for industrialized countries. According to Cohen, the "proposition that the theories of risk society and ecological modernisation are positioned in opposition to one another provides the foundation for the two-dimensional typology" (110) However, this interpretation raises some problems. As we will see later in this chapter, for Beck (1992), the theory of RS arises not from the absence but from the presence of a strong commitment to scientific rationality. It is therefore problematic to try to distinguish EM and RS from this point. The view that sees the two as opposing perspectives is not as satisfying as it seems.

Mol and Spaargaren (1993) present their arguments in the form of a provisional hypothesis, arguing that RS simply seems to contradict the theory of EM. Thus, they seem reluctant to take their own arguments literally. Moreover, at the end of their article, they concede that when analyzed in the context of science and technology, there is not only tension between RS and EM, but also, in some sense, a relationship of complementarity. For Mol and Spaargaren (456), one of the aspects that RS theory contributes to EM theory is that it offers an analysis of the reflexive nature of science and technology.

The pessimistic approach, given by Beck (1992), toward science and technology would seem sufficient to set him in opposition to any approach that recognizes their importance for a sustainable future. However, we will try to show that this can be a hasty interpretation. The theory of RS is fundamental for us to assess some existing contradictions, both in the discourse and the theory of EM as well as in the concept of sustainability. So, to analyze the possible problems and potentialities linking sustainability, science, and technology, the RS perspective becomes a fundamental approach. Next, we will examine the work of Beck (1992) and Giddens (1990), focusing on these issues.

THE COMING OF RISK SOCIETY

For Beck (1992, 21), risks arise in response to the uncertainties and insecurities produced by modernization. Dangers and risks are part of human history, and for Beck they are an inherent aspect of human action. This would seem to make the category of risk an imprecise means to differentiate social epochs. However, in *Ecological Politics in an Age of Risk*, Beck (1995a) provides a distinction between the specific risks of nonmodern cultures and those of the industrial society and RS phases. In this work, Beck makes two basic distinctions regarding the question of risk and types of society, the first distinguishing nonmodern cultures from modern ones, and the second between the two phases of modernity (industrial society and RS). According to Beck, premodern cultures or societies did not specifically face risks, but rather dangers. The distinction between one and the other basically resides in their origin. Risks imply a choice. Dangers differ essentially from risks, since they are not the result of decisions based on technoeconomic opportunities. Risks have their origin in the threats produced by modernization.

It is possible to see many similarities between the threats that were present at the time of the emergence of industrial society, in the nineteenth and twentieth centuries, and those that characterize the present period. Although modernization is associated with social and environmental risks at different stages of modernity, the relationship between the two takes place on different bases. Beck identifies three basic criteria for distinguishing the risks of the present from those arising from industrialization in the early nineteenth century, characteristic of industrial society. There are three factors that clearly characterize this difference. The latest risks have no clear boundaries, social or spatial; nor can they be readily subjected to the rules of causality, fault, and liability; nor is it possible to counter them with the usual safety strategies (1995a, 2). These difficulties arise because the most serious and complex risks we must deal with are global in scope. Some environmental risks fall into this category. Beck (2000) points to three different types of global risks: (a) those arising from prosperity and associated with techno-industrial development, (b) those arising from poverty, and (c) those arising from the use of weapons of mass destruction in exceptional circumstances.

The risks involved in the industrial society were related to the creation and distribution of wealth. At stake in the industrial society was the struggle between capital and labor for the fruits and benefits of an industrial system focused on the creation of material goods and services. In RS, a very different process is taking place. The main struggle is not over access to and distribution of these goods, but over the power to prevent or distribute the evils arising from modernization itself. As Beck makes clear, at the beginning of RS, in addition to the social conflicts associated with the distribution of wealth, there are at the same time the conflicts arising from environmental risks due to technoscientific development. These conflicts revolve around the generation, definition, and distribution of these environmental risks (1992, 19). Thus, the social risks associated with the conflicts between capital and labor are not replaced by environmental risks, but the two overlap in RS and in some ways rival each other in the importance of the disputes on the public agenda. In several passages, Beck suggests that social risks (unemployment, poverty, etc.) override the debate about environmental risks and cause the political system to address the usual responses to these issues. However, as environmental risks may worsen and lead to increased public concern, the debate over environmental risks may gain prominence in the public agenda.

In addition, the new risks involve a special process of victimization. On the RS site, class and risk position may not coincide. Although Beck (1992, 35) acknowledges that the state of poverty determines the type of risks people face, the economic criterion is not sufficient to explain the dynamics of risk in RS. As risks tend to intensify, they create what Beck calls the "end of the other." The consequential ecological risks can lead to situations where no norm can confine them to particular social groups (black/white, rich/poor, male/female, etc.). Whereas human suffering in the past was processed from the category of "others," which consisted of human groups that were inevitably the focus of violence and social exclusion (blacks, women, Jews, exiles, dissidents, etc.), this distance between "us" and "others" tends to disappear in the face of the new far-reaching ecological risks. As Beck (1992, 36) notes, poverty might be marginalized in IS, but that hardly applies to the major risks that arise within RS.

In this sense, RS creates a kind of negative equality. High-consequence ecological risks become democratic because they do not follow any social dividing line that we know of; they can literally break down all the barriers that separate diverse social groups in societies. People become equal not because of the rights or benefits they achieve, but because of the environmental damage they share. Global pollution, unlike poverty, does not follow clear social boundaries, although as Beck (1992) suggests, it can in certain cases. And in those cases, there tends to be a convergence between social and environmental conflicts. Thus, what emerges in RS are communities of danger (Beck 1992, 47). In these cases, the common denominator that demarcates the "others" is not a particular social boundary (a territory), but the simple condition of being exposed to the same risks and dangers. One might think there is a logic behind this that determines who wins and who loses. However, as Offe (1992) has shown, RS provides space for negative sum game situations in which everyone causes harm to themselves and others. The risk producers themselves may suffer consequences that are not the same as those of the usual victims but still involve some kind of loss. RS thus assumes there is a zero-sum game with environmental risks in certain situations. Some groups may lose more than others, but in general everyone has something to lose from the resulting consequences.

Another differential of the new ecological risks concerns their scope. The new risks have a universalizing and globalizing trend, accompanying the globalization of industrial production and becoming independent of the place where they are produced. For Beck (1992), the food chain unites practically everyone on the planet. For this reason, RS is also a world society and the "danger community" transcends not only social but also political and geographic borders. So the community of danger is delimited by exposure to risk; therefore, it might not be out of place to consider in this context that in certain situations, as in the case of climate change, the "community" in question becomes a "global society."

A key aspect of these risks is their catastrophic dimension. Beck repeatedly warns of the catastrophic effects that technology can have in the nuclear, genetic, and chemical ages. The term RS itself refers to a state of emergency of societies facing large-scale destruction of the planet (1995a, 67). This thesis projects a globality into the idea of RS. In his early works, Beck (1992) used the term RS without adjectives, emphasizing its global character only in certain passages. This happened because his theses initially gained force when applied to the context of countries such as Germany, his country of origin. More recently, however, the author has used the term "global risk society" instead. For Beck, the global nature of RS is illustrated in works such as *World Risk Society* (1999) and *What Is Globalization?* (2000). Although Beck initially relates the social change of RS to countries such as Germany, some statements suggest that the change is linked to issues of a global order. This does not necessarily imply a contradiction in his social theory, for, in Beck's assessment, the reaction to environmental risks tends to produce precisely this contradiction. There is a mismatch between the globalizing features of the new risks and the institutional responses to them. As Matten notes, while governments seek to gain control over their territory, under the conditions of the RS, ecological globalization presents itself as something that is happens "beyond these national borders, governmental control over society shrinks gradually as well" (2004, 389). This contradiction is precisely one of the processes that fosters the ecological crisis (RS). Old answers to new problems tend to create the conditions for aggravation of those very problems.

The logic governing ownership of material goods is not the same when we talk about risks. This is because risk never takes a clear concrete and material form. This is not only because risks refer to potential future threats, but also because they often involve a process of expropriation of the senses that makes them imperceptible. Scientific knowledge plays a central role in this context. People are increasingly dependent on science and its representatives (research institutes, scientists, and specialists) to understand what is happening to them and to nature. Science becomes a mediator through which risks are perceived in a process in which the perception of environmental reality (risk) is partially hijacked. As Beck (1992, 27) argues, the latest risks elude our ability to perceive them. Many of the potential hazards posed by techno-scientific development are not visible or perceptible to victims because our ability to perceive these risks is very limited. Therefore, in his opinion, to perceive these risks, we need the "sensory organs" of science. We must make use of their tools (theories, experiments, measurements, etc.) to make visible what is beyond our ability to perceive.

Risk and risk perception are therefore intertwined, and perception in this case is largely about scientific perception. In this way, scientific knowledge becomes essential to define even who is or is not at risk, who are the potential and actual victims. Simple exposure to environmental risks also does not explain the conflicts of the RS, because victims must recognize their exposure for these conflicts to arise. This means the groups that tend to be affected by risks are those who have access to knowledge or produce it in some way. Thus, in the RS, the condition of victim is not exclusively determined by exposure to risk but involves a re-elaboration of experience through scientific knowledge that is acquired about the risks. Scientific rationality develops, in many cases, in a conflicting interaction with lay knowledge.

SCIENTIFIC RATIONALITY IN QUESTION

What distances the work of Beck from the concepts of SD and EM, at least initially, is the interconnection he establishes between science, technology, and environmental risks. As we have seen, the very acceptance of the existence of dangers requires the "sensory organs" of science, capable of making the threats of modernization visible and interpretable. However, according to

Beck (1992), there is a series of factors that can generate misunderstandings in scientific perception and that contribute, in turn, to the creation, legitimization, and proliferation of risks. In several passages of his works, Beck affirms that science is unable to recognize the risks and problems it produces, hence the controversy between the theory of RS and the concepts of SD and EM. For example, in *Risk Society* there are several passages in which Beck blames the sciences for the risks they produce. In one of them, he states that science is unable to respond to new risks because it is involved in the creation and growth of environmental risks (1992, 59). Somewhat further in this work, he sees science as the protector of global contamination of humans and nature (70). On the other hand, and somewhat paradoxically, alongside these passages are some statements that show that his vision seems less categorical. In them, science is also presented as a friend or ally to those who want to find an answer to environmental risks. He acknowledges that our perception of risk depends simultaneously on a scientific and social construction. Thus, he asserts that science is one of the causes of new risks, it is also "the medium of definition and the source of solutions to risks" (155).

The relationship between science and the new risks of modernization is not, as one might think, only negative, in which science would present itself only as a major "source of problems." Science is also recognized by Beck (1992), as the last quote shows, as a condition through which the risks of modernization can be recognized. Thus, science can also be seen as a "source of solutions," or at least as an important and necessary step in solving environmental problems. The position of science in RS is thus characterized by ambivalence. Science is the source of certain environmental risks, but it is found simultaneously in the perception of these risks and also in the solutions that are prescribed for them. This shows that it may be reductionist to characterize Beck as a clearly pessimistic and critical author of science and technology. For him, science occupies an ambivalent position: Science is both the problem and the solution simultaneously.

The institutional connection of science to new risks arises in part from the existing pact between science, technological innovation, and productivity. New risks no longer arise in a context of poverty but rather in one of prosperity, in which science provides a major stimulus to economic growth through technology. New risks of great magnitude are thus due to a victory of industrial society: its potential for overproduction. In this context, science and modern technologies occupy a central place. Having become a factor in promoting this hyperproductivity, science is directly involved in the production of risks through high-risk technologies. This is one of the contributions of science to the production and legitimization of modern risks. In the drive to increase productivity, ecological risks are often overlooked. For Beck, science tends to be guided by economic interests, and "in the effort to increase productivity, the associated risks have always been and still are being neglected" (1992, 60).

By allowing itself to be instrumentalized by the economic sphere, seeking higher productivity at any cost, science develops a systematic blindness to environmental risks. This shows that we cannot neglect the intertwining of technoscientific development with modern economic demands and the impact of this relationship on environmental risks. The problem occurs more frequently in the application of modern technologies, although it is also related to what many consider existing flaws in current scientific theories. Let us take just one example. Many biotechnologies are being created on the basis of assumptions contained in modern genetic sciences. According to several authors, the latter has detached itself from modern biology and opted for epistemological reductionism, warranted only by the financial benefits such technologies bring in the economic sphere. Modern genetics has abandoned the traditional biological perspective, which focused on organisms and species, to reduce the dynamics of the natural world to genes.[3]

In addition to the economic role that scientific and technological development can play, two other aspects should be emphasized regarding the link between science and the modern economy. Political and economic interests do not intersect with scientific and technological development in an anticipatory way in the form of investment in the production of technologies, but rather arise when the risks are socially recognized. In these cases, social recognition of risks simultaneously leads to recognition of the interests of the modernizing actors who produced them or who seek to benefit from them in some way (Beck, 1992, 27). Moreover, political and economic interests can shape and influence the way we perceive risk. These interests permeate the cause-and-effect relationships established between human actions and environmental impacts. They are part of the "social context" in which risks are assessed and can influence how we respond to them.

Another factor linking science, economics, and risk is the process of specialization of scientific knowledge. Although specialization is in part a specific path to science, it is also associated with the risks of modernization and the modern economy.[4] For Beck, a condition for the emergence of environmental risk in the context of RS is the existence of "specialization of cognitive practice" (1992, 178). The greater the existing specialization, Beck argues, the lower our ability to deal with the risks we produce. The argument that scientific specialization simultaneously ignores and legitimizes the production of risks recurs throughout his work. It is important to point out that for Beck, specialization in knowledge is a mirror image of specialization in the productive sphere; higher productivity goes hand in hand with a finer division of labor and knowledge. It should be noted, then, that the division Beck refers to is not only to the division between fields or disciplines, but to a number of divisions that have become established in modernity and that extend beyond the scientific realm. These distinctions include the separation between theory and practice, facts and values, science and ethics, and, what seems essential to Beck's analysis, "the institutions that support these divisions" (70). In this context, science is going through a radical process of reflexivity. His theses on the transition from simple to reflective (or secondary) scientization are based on the relationship between scientific practice and the public sphere. Beck investigates the legitimacy and public trust on which technoscientific development depends. This legitimacy of scientific practice is, according to him, called into question by the circumstances in which a social explosion of risk occurs, an issue to which we will return later.

In *RS*, for example, Beck notes that the criticism of science in the context of RS stems from the failure of scientific rationality to deal with technological risks. This failure does not emanate from individual scientists or disciplines, but from the institutional approach of the sciences to risk (1992, 59). One of the most controversial points in *RS*'s thesis lies in the definition of science itself and its relationship to the risks of modernization. In addition to the fact that the social context of knowledge may promote the emergence of risks, the constitutive principles of scientific discourse may also contribute. Although the definition of scientific rationality and the principles and values that belong to it may be contested, it would be difficult to deny that the identity and authority of modern science rests on pillars such as (a) distinction between fact and value, (b) certainty, (c) experimentation, (d) causality, (e) distinction between theory and practice (or theory and application), (f) methodological skepticism, (g) specialization, and (h) distinction between scientific and lay knowledge. In Beck's (1992) work, some of these pillars are associated with negligence and the production of modernization risks.

The fact that science recognizes only those relationships between causes and effects that can be demonstrated theoretically and empirically has serious consequences for attempts to respond to ecological risks. The emphasis on causality leads to political neglect of a range of risks whose relationships cannot be scientifically established and subsequently socially acknowledged. This is because of the complexity that these risks entail. For standard science and for the political and legal system that uses it, threats and hazards are recognized as risks only if they allow a causal explanation between modernization processes, effects on the environment, and people. In the RS, there is a contradiction between actual environmental degradation, which occurs in uneven and diverse ways across the planet, and rigid criteria for establishing scientific causality, a type of causality that ultimately contributes to the proliferation of risks in the name of the quality of scientific research.

In short, insistence on scientific "purity" leads to "dirt" in the environment. This is because in the case of many modern risks, especially ecological risks with large consequences, it is not possible to establish a direct and clear relationship between causes and effects. In this case, insisting on a close relationship between environmental impacts and modernization leads to a failure to recognize the threats posed by industrial pollution in more subtle ways. The causal link between modernization and risk is difficult or often impossible to prove. Thus, although it is one of the central elements of scientific rationality, it is fundamentally inadequate to assess the risks of modernization. Finally, this inadequacy has implications for the legal system itself. In the case of environmental pollution, the impossibility of proving causality, and thus responsibility, reduces the polluter pays principle to a normative economic ideal with no effective practical implications, according to Beck (1992, 63).

This ambiguity of scientific principles in relation to the risks of modernization leads to the question of skepticism that pervades scientific logic. Methodological skepticism, which drives the constant refutation of scientific hypotheses and evidence, has become a process inherent in the dynamics of the scientific field in modernity. Thus, scientific development itself involves the "creation of uncertainties." The traditional development of science depends on the formulation of criticisms, contradictory results, different approaches, and so on. However, when this relates to the risks of modernization, the result is to invalidate the various discourses that have been built around these risks. The complexity of these problems can lead to a multiplicity of causal interpretations in which each discourse cancels out the other (Beck 1987, 157). In this scenario, no "truth" or "consensus" emerges about the risks under debate, but rather a general doubt that paralyzes political discourse and action.

Another problem concerns scientific experiments. An important aspect of the sustainability concept is the establishment of "tolerable" or "acceptable" levels of pollution. Any sustainability policy requires the adoption of measurement methods to assess the extent to which our interventions in the environment promote environmental degradation or ensure its preservation. At the same time, there is a contemporary tendency to assume that these acceptable levels can be determined by science. In general, science positions itself as an authority capable of setting the safety level of new technological systems and the tolerance level for pollution of ecological systems. However, according to Beck (1992), science cannot answer the question "How much security is sufficiently secure?" on its own. Science is always guided by a "state of the art," which is always provisional in terms of safety. Now, technologies must be tested outside the laboratory so that their safety can be evaluated. Therefore, any prediction of safety standards for pollution must be seen as provisional, which makes the desired safety, as Beck points out, only a "probable safety"—not to mention that views on risks and safety are permeated by cultural values.

All of this invites discussion of modern science. First, safety considerations break down the distinction between facts and values on which standard scientific discourse is based. Second, the ideal conditions for experimentation and control are no longer present, as the establishment of tolerance values depends on the direct application of technologies in society. In this case, there is a kind of social experimentalism that would transform RS into a "laboratory society." In trying to set acceptable levels, we face the problem of knowing in advance whether or not we are contaminating people and the environment. With toxic substances, for example, we would need to know in advance what is toxic and what is not. We should be aware of the perverse consequences that the emission of certain substances can bring. However, in the case of new technologies, this is not always possible.

Take the case of nuclear energy. Nuclear reactors cannot be tested inside a laboratory. The development of this type of technology has been possible only with its direct implementation in society, in a process in which there is no clear line demarcating the transition between "laboratory" and "application." Here, no clear distinction can be made between "theory" and "practice," nor even between "fact" and "value," since the scientific experience in question has direct political implications. In the RS, these processes are simultaneous. "Science has itself abolished," says Beck, "the boundary between laboratory and society" (1992, 108). In this context, the very meaning of experimentation seems to lose its more usual sense, since it can only be performed simultaneously with the direct application in society.

The problem also exists with technological innovations, where safety conditions cannot ultimately be fully tested. Attempts are made to test the safety of certain products through animal testing, but the information these tests can provide is in many cases uncertain. Reactions vary from animal to animal and bear little resemblance to human reactions. Since it is not possible to pretest such technologies and make a credible judgement, the substances are simply released into the environment. In many cases, Beck (1992) says, it is virtually impossible to obtain reliable knowledge about the safety of substances or technologies before they are introduced to or disseminated in society. The problem lies not only in the lack of demarcation between "society" and "laboratory" that occurs in this process, but also in the removal of conditions normally associated with the idea of experimentation. When new technologies are introduced, scientific experiments simply do not take place because people's reactions are not systematically studied and recorded.

The distinction between fact and value is also important when considering technical security conditions. Would it be possible for us, then, to have a purely technical and scientific assessment of safety risks and standards? The answer Beck gives is negative. Our assessments of environmental or technological risks always involve some kind of normative judgment. The environmental movement's resistance to the risks posed by new technologies is described by many analysts as irrational. Beck (1992) does not deny that contemporary environmental thinking contains some contradictions, but technicians and scientists would also put forward their fallacies if they made a rigid distinction between "experts" and "nonspecialists" in risks. Such a distinction could be detrimental to the discussion of conflicts between experts and the public because it reduces the conflict to a mere information problem on either side. In this view, the information deficit would affect the environmental movement and the lay public, rather than, say, technicians and scientists. For Beck, "there is no specialist in risk" (29).

For Beck (1992), risk statements contain views about how we want to live. They thus conceal arguments about life choices, about how people, especially the public exposed to these risks, want to live. To the extent that such decisions are moral, they cannot be addressed by experts. There are no "risk experts" because, in a sense, there are no "moral experts" who can dictate life choices to others. Scientific statements about risk, if unaware of this normative background, are trapped in a kind of technocratic view of risk that obscures the moral issues at stake, issues in which the public might be interested. This suggests that deliberations about risk should be not technical but democratic deliberations. Disputes about risks are moral disputes and, for that very reason, political disputes. For to the extent that the public evaluates risks differently, a challenge to decision making arises in this scenario.

Beck (1992) is not the only voice pointing out the ambivalence of contemporary technoscientific development. Krohn and Weyer (1994, 173), who largely support Beck's critique, argue that modern science and technology tend to promote what they call "real-life experimentation." Poel (2017) addresses the same questions by viewing the impact of new technologies as a kind of social experiment that puts society in the state of a "laboratory." According to Poel, new technologies are a kind of social experiment because they are a kind of innovation that takes place (a) *in* society, (b) *on* society, and (c) *by* society. Regarding point (a), it can be said that new technologies do not emerge in a limited setting, in the form of a laboratory, but develop within the framework of society itself or a part of it. Regarding point (b), it can be said that new technologies are an experiment for society in that it may suffer from the negative effects that arise, "but also in terms of their social consequences, risks, social embeddedness, and normative dimensions" (64). And point (c) occurs because these innovations are driven not by a single experimenter but by a set of actors who participate in the new innovations. What Beck's RS theory requires of the emerging risks of science and technology is, in a sense, a regulatory framework for technoscientific development.

Under these circumstances, modern science would extend the research process beyond the confines of the laboratory to include other institutions

and society as a whole. To the extent that the values inherent in technological innovation are elided and the values that come from those who wish to
oppose such change are rejected, the problem of the legitimacy of technological change arises. Given the different values underlying conflicts over new
risks, one of the central problems of the RS theory's approach to science
concerns its social legitimacy. These problematic aspects related to modern
science and ecological risks are part of a contradictory dynamic in which RS
is embedded. Its contradictory character is represented by the concepts of
organized irresponsibility, definitional relations, and the social explosion of
risk, which we will analyze below.

COLLECTIVE IRRESPONSIBILITY IN RISK SOCIETY

Beck suggests that many decisions that promote risk production tend to be
scientific decisions insofar as they are guided by scientific criteria. In many
of these cases, however, scientific decisions are also political decisions, since
in this interpretation these decisions are assumed to require the authority of
political power. These risks are scientific, technical, and also political. This
means that science does not exert its influence on risk perception in isolation
from the political process, but rather the opposite; it exerts its influence to the
extent that it brings its own criteria to bear on the political process. Thus, the
failure of science to assess risk is also a political failure and thus a triggering
factor for political crises.

Mistakes in science and technology, in the production and legitimization
of the risks of modernization, lead to a contradictory institutional process
in RS. This institutional contradiction arises because the failure to confront
these risks can lead to a contradiction within the institutions responsible for
dealing with them. Thus, in a sense, the focus of RS theory is on the generation of an institutional crisis in modern societies. RS provides a diffuse and
complex framework in which the risks generated are not attributable to any-
one. This inability to account for the production of these risks leads to what
Beck (1995a) calls organized irresponsibility. This concept seeks to translate
the latent institutional contradiction, in which threats are produced, but no
one is responsible for them. This institutional crisis can be seen as a crisis of
responsibility that, according to Beck (1995b, 109), forces us to rethink how
responsibility can be attributed in the context of RS. This leads us to a revision of its own regulation in society.

Organized irresponsibility develops from two contrasting historical developments of IS: on the one hand, the emergence of threats and risks from
industrialism itself, and, on the other hand, the creation of bureaucratic
security standards and systems in response to the former. IS, according to

Beck (1992), created antidotes (security systems) for the "poisons" (risks) it generated. In IS, then, safety and risk are two sides of the same coin. This shows that the characteristic risks of modern societies emerge in a specific political-institutional framework against the background of a social pact in which institutions take responsibility for controlling the risks generated by industrialism.

The existence of a welfare state is a fundamental institutional condition for the idea of organized irresponsibility to have any meaning at all. This crisis of responsibility, according to Beck (1992), exists in countries where the state has taken on the task of dealing with the new risks that industrialism has produced.[5] The modern welfare state proceeds in two ways in its confrontation with the risks posed by industrialism. The first step is to make the threats a "calculable" phenomenon; that is, risks. In order to deal with the potential threats posed by industrialism, they must first be made calculable. Thus, the threats arising from industrialization itself (e.g., industrial accidents, unemployment) are considered calculable and predictable. Thanks to this calculation of threats, the state can then respond in the form of compensation and prevention.

For Beck (1992), this form of institutional control and responsibility worked relatively well until the emergence of "normal accidents."[6] In *Normal Accidents*, Charles Perrow (1984) warns that complex technologies such as nuclear power, biotechnology, and the chemical industry have an imponderable common element. Regardless of what conventional safety strategies are employed, some form of accident inevitably occurs. Perrow (1984) calls this a normal or system accident.[7] The failures that trigger these accidents cannot be traced to anything in particular but arise from the complexity of each technological system itself. While some kind of learning about their failures was possible in early twentieth-century systems, such learning is possible in more recent transformational systems that combine the features of interactive complexity and tight coupling. In the author's view, there is a decreasing curve in the learning about failures of these technological systems over time.[8]

When accidents involving these technologies are "normal," the strategies and responses developed by state institutions and other responsible actors are called into question. These normal accidents confront institutions with the question of who bears responsibility for the consequences of their actions and put pressure on institutions to ensure higher and more reliable safety standards. In the case of nuclear, biotechnological, and chemical accidents, however, the usual forms of control and responsibility are not satisfactory. The high costs outweigh the possibility of compensation, and the complexity of the case also makes successful preventive measures difficult. Organized irresponsibility, then, refers to a normalization of danger, the process by which hazards become acceptable or tolerable. It helps to explain how and why modern institutions are forced to know the reality of dangers but at the same time deny responsibility for them. Organized irresponsibility is paradoxical because it points to a situation in which, as environmental degradation increases and legal and political responses to it accumulate, no one is held accountable for what is happening (Beck, 1999, 149).

This accountability crisis is fueled by emerging definitional disputes. Beck refers to the definition of specific rules, institutions, and capacities that structure the identification and assessment of risk in a particular cultural context. Such definitions constitute the legal, epistemological, and cultural matrix through which risk policy is conducted (Beck 1999, 149). These definitional relationships are formed by four types of questions: (a) Who should define the existence or severity of hazards? Who should be held responsible for their existence? Should that responsibility lie with those who have produced them, with those who have benefited from them, or with those who are affected by them and tend to lose out? (b) What kind of knowledge should guide the assessment of the causes and magnitude of hazards (scientific/lay)? How does one deal with ignorance and uncertainty about risks? (c) What evidence is sufficient to identify hazards? How can this be possible in a situation where knowledge of hazards is always challenged? (d) How should victims be compensated, and who determines the forms of hazard limitation, control, and regulation? (1999, 149).

This is where the political struggle to define the magnitude and urgency of the risks lies. Since the dangers created by modernization are neither obvious nor consensual, there is a real struggle to establish what the "real" threats are and their potential social impact. According to Beck (1999, 46), definitional relationships are developed at three basic levels: (a) the production of knowledge (definitions) about risks, (b) the dissemination of that knowledge (media), and (c) the reception of and response to that knowledge. These definitional disputes affect information, data, evidence, and relevant knowledge about environmental hazards. So does the ability to name culprits and find possible responses. They are also what may (or may not) lead to a situation of organized irresponsibility.

This brings us to the role the contemporary legal system plays in the production of organized irresponsibility. Issues such as the burden of proof or the legal principle of individual culpability are related to the legal system. The first defines who must prove the existence or severity of existing hazards; the second, how culpability must be attributed. As Beck (1992) notes, in a context of universal pollution, where everyone contributes to its occurrence to varying degrees, it is impossible to attribute it exclusively to a particular individual. Indeed, this pollution is often caused by corporations and organizations rather than by individuals. The diffuse nature of modern pollution thus weakens the effectiveness of the principle of individual culpability. It is not even possible to attribute accidents associated with high-risk technologies simply to "operator error," since the fault lies in the complexity of the technological system itself, as pointed out by authors such as Perrow (1984). It is difficult to find someone guilty in a framework where everyone (producers, consumers, government, etc.) seems to be involved. The legal system itself is the cause of the emergence of environmental risks and thus of the promotion of a situation of organized irresponsibility. As Beck notes, definitional relations make the legal system complicit in pervasive environmental pollution (1995b, 134).

This picture of organized irresponsibility is not static, however. It is affected when a social explosion of risk occurs. The concept of the social explosion of risk is used by Beck (1995a) as a metaphor to explain the politicizing effect of conflicts over the definition of risk. Risks of great magnitude and the uncertainties they generate shape a dynamic of political change that, according to Beck, undermines state bureaucracies and challenges the dominance of science (1999, 250). With the social explosion of risk, strategies for concealing institutional contradictions (organized irresponsibility) become the focus of public attention. In this political scenario, the foundations supporting the normalization of risk are shaken, and the political order undergoes a transformation (Beck 1992, 76).

The social explosion of risk thus means the preliminary unveiling of the hidden contradictions and strategies behind organized irresponsibility. For this to occur, first, organized irresponsibility and the dangers it poses must threaten desired social values. Second, the existence of a welfare state seems to be an important precondition. Third, there must be conflict between those who gain and those who lose from the risks generated. In addition, an independent press is essential to amplify conflict. Finally, the environmental movement plays an important role in highlighting and exposing the contradictions of the system. Moreover, it should be emphasized that the strength of the environmental movement may be linked to, and even dependent on, the institutional contradictions generated by organized irresponsibility. In all of this, Beck sees in the social recognition of risk a political potential that could lead to a reordering of power.

SUB-POLITICS AND ECOLOGICAL DEMOCRACY

In RS, the meaning of politics is transformed. The political alternatives for dealing with the dilemmas presented above are hidden in these changes. The question of knowledge is central to RS, so politics turns to science and knowledge in general. As Goldblatt notes, with the decline of class politics in RS, the importance of the politics of knowledge and the role of those who produce, disseminate, and interpret that knowledge increases. Thus, in his view, "it is not surprising to find that Beck places the discourse and practice of science at the heart of the politics of the risk society" (1996, 161). For as we have seen, science is involved in the creation of environmental risks, in our perceptions of those risks, and also in the possible responses we can make to them. Knowledge plays an important role at all levels associated with risk conflicts.

The politics that RS sets in motion is referred to by Beck (1992) as sub-politics. The term sub-politics is no longer equated with the traditional institutions of political life (parliament, parties, state), but refers to a new kind of political culture that operates outside or beyond these institutions. Beck (1992, 194) adds to the definition by noting that the emergence of sub-politics is linked to political modernization through civil rights. He points out that the centers of sub-politics are created and stabilized simultaneously with the establishment of basic rights. Beck defines sub-politics as "politics outside and beyond the representative institutions of the political system of nation-states" (1999, 39). It emerges with the social recognition of risks and the political consequences that result from this process. It is an expression of a process of political self-organization that emerges with the social explosion of risk. Sub-politics, as Beck (1999, 39) notes, tends to transform into a kind of direct politics in which society is shaped from below. It is a direct politics because, as we can see, citizens and civil society organizations now have a more direct influence on decisions about risk. And in this scenario, the relationships for defining risks, which are also power relationships, are changing.

We can say that sub-politics arises where government environmental policy is trapped in naturalistic approaches that disregard public perceptions, norms, and values. It can also arise when the same policy is paralyzed by internal conflicts of interest or when it proves too rigid and bureaucratic. The arenas of sub-politics are diverse and always lie outside traditional political institutions, but still possess decision-making power and influence in structuring contemporary societies. They include, as Mol, Hogenboom, and Spaargaren (2001, 98) elaborate, the economy, corporations, unions, mass media, the legal system, the family, academia, and social movements. They are political processes in which civil society participates and which influence political decisions about environmental risks but do not originate and develop in institutions such as political parties, parliament, or the executive branches of government.[9]

One way to respond to the dilemmas of organized irresponsibility analyzed above would be to strengthen and expand the conditions for the functioning of this sub-politics. Several strategies could strengthen it, making it easier to address the problems that arise from organized irresponsibility. Action would be needed in three key areas that are important for a modern sub-politics to thrive: (a) a strong and independent legal system, (b) free and critical means of communication, and (c) a process of self-criticism based on various forms of knowledge about risk (Beck 1992a, 234). In addition to strengthening these arenas of sub-politics, Beck envisions two strategies for dealing with the contradictions inherent in RS, exposing the imperfections of technocratic notions of security and danger and creating space for opposition to dominant views. One of the strategies is to denounce the existing deficiencies in the concepts of risk and security and to highlight the existing deficit in cultural acceptance. This is what Beck calls the "denormalization of acceptance" (1995a, 173). After challenging definitions of security of the bureaucratic elites, a second strategy is to fight for a broader concept of security. Technological issues concerning the "worst possible scenario" should not be excluded from this concept, the expansion of which could be achieved by breaking the social monopoly of the groups that define the risks. As can be seen, these strategies aim to change the existing power relations in risk definition relationships. Most of them create better conditions for civil society to influence risk definition more directly, while threatening the monopoly that technocrats have on risk definition.

The change in the defining relationships is fundamental and intertwined with these two strategies. As we have seen, the way in which disputes over definition are processed in industrial society equates economic and technological change with social progress. This means that risks and dangers are discussed only after certain technological and economic practices have already taken root in social life. Even more perverse, the burden of proof for adverse effects rests with potential victims, not producers. One way to get ahead of the dangers would be to reverse this situation so that the burden of proof is not on the actual or potential victims, but on the actual or potential perpetrators of environmental degradation. Scientists, companies, and technicians must prove that their initiatives will not lead to hazards, not those who will be harmed by such initiatives. In addition, the forms of deliberation should be rethought.

There are also several possibilities with respect to the principle of individual culpability and the forms of attribution and responsibility. These political strategies are part of what Beck calls a differentiated politics of RS, which is not a politics "from above" but one based on self-politicization by the already existing sub-politics. Just as for Marx capitalism produces its gravedigger, for Beck organized irresponsibility stimulates a new political culture that can turn against the perverse consequences that the first produces. However, although this parallel can be drawn between RS and Marxism, there is no "ecological proletariat" in RS that can constitute itself as the political subject of a change dictated by history. For Beck, the substitute for the "ecological proletariat" is the political reflexivity that the environmental conflict generated by organized irresponsibility tends to produce. For Beck (1995a, 3), the political subject in class society corresponds to political reflexivity in RS. At the same time, there is no obvious direction toward which this kind of environmental conflict tends. What it does produce, however, is a set of political possibilities that those who resist environmental technocratism can ultimately exploit.

Ultimately, what is at stake is the construction of an ecological or reflexive democracy. Beck does not give a clear definition of what he means by this ecological democracy, but two main points can be noted. First, ecological democracy is based on this sub-political movement and thus represents a more participatory form of democracy exercised through alternative means to the usual channels of politics (parliament, parties). The political possibilities offered by the environmental conflict triggered by organized irresponsibility open a space for the political mobilization of civil society. This does not mean that direct democracy is emerging in RS, but that representative liberal democracy tends to be influenced by political processes that in some ways express a more direct kind of political participation. The emergence of sub-politics is accompanied by a distrust of the usual democratic institutions. Second, an ecological democracy would break with majority rule and be guided by a constitution capable of learning (or reflective). The first idea is taken from Claus Offe and the second from Ulrich Preuss. Democratic majority rule, in these circumstances, would only make the resolution of social conflicts more difficult, as well as reducing complex environmental issues to terms of mutually exclusive alternatives ("yes" and "no"). A reflexive constitution, in turn, would have the role of keeping the future open and would make room for the veto power of minorities.

MODERNITY AND THE END OF NATURE

Giddens's conception of modernity bears many similarities to Beck's theory of social risk. In *Modernity and Self-Identity*, Giddens concedes that it is quite accurate to characterize modernity, as Beck does, as a "risk society" (Giddens 1991, 28). Giddens (1990) uses various expressions to examine the environmental changes brought about by modernity; created environment, socialized nature, end of nature, and ecological risk are some of them. Although some of these terms have similar meanings, the use of some of them varies according to the development of their ideas. Giddens offers a reading of current environmental changes from his own understanding of modernity. In *The Consequences of Modernity*, he characterizes modernity as the lifestyle or social organization that emerged in Europe around the seventeenth century

(1990, 1). For Giddens, modernity is unique in terms of the dynamics, scope, and institutional nature of the changes it sets in motion.[10]

In *Modernity and Self-Identity*, Giddens states that modernity expresses an orientation toward control, an orientation that suggests an attempt to subordinate the external world to human domination, and "one thing control means is the subordination of nature to human purposes" (1991, 144). For Giddens, modernity tends toward the domination of nature, and to some extent its relationship with nature can be seen in this light, although the approach he proposes differs from that of those who seek to understand this relationship through a unilateral domination of instrumental rationality. This control occurs through two axes that integrate modernity itself: capitalism and industrialism. In works such as *The Nation-State and Violence*, Giddens (1985) does assert that industrialism and capitalism are the origin of the existing environmental changes in modernity, changes he addresses with the concept of the "created environment," but there is no clear indication of how each of these dimensions relates to these changes. But how then does Giddens understand capitalism and industrialism? And what relationships can be established between these dimensions of modernity and environmental change?

Drawing on Marx's and Weber's interpretations of capitalism, Giddens develops a vision that attempts to transcend both. First, he distinguishes between capitalism and capitalist society. Capitalism, in his view, can be used to refer to a set of economic activities that are isolated from political activities. These economic activities are based on the existence of private property and require financial accounting that organizes the balance of costs, profits, and opportunities for reinvestment. The word can also be used as a synonym for capitalist society, but in this second meaning it encompasses a broader range of issues such as the following: (1) the economic activities that form the primary basis for the production of goods and services and on which a portion of the population depends; (2) the isolation of the economic sphere from the political sphere, which does not necessarily mean the absence of state interference, but may imply it; (3) the establishment of private property as an institution (where private property means the control of capital by nonstate entities); (4) the role of the state, which is influenced by the process of capital accumulation; (5) the existence of a nation-state that guarantees borders. Thus, a capitalist society presupposes the existence of a nation-state, whereas capitalism as an economic system does not. (Giddens 1985, 136–37).

Capitalism refers to the way economic relations are institutionalized. It refers to the organization of the market under the conditions of modernity. Industrialism, on the other hand, refers to the rational organization of the production process. It refers to the types of resources used in the labor process and the way production activities are organized. Industrialism, in turn, is characterized by (1) the inanimate use of material energy resources in the process of production and circulation of goods; (2) the mechanization of production and the economic process in general; and (3) the predominance of manufacturing industry, not in the sense of nonagricultural goods, but as indicating a way of organizing production based on the conjunction of points (1) and (2). Industrialism is also characterized by (4) a spatial centralization of production activity (Giddens 1985, 136–37). These distinctions are primarily of analytical interest. In most modern capitalist societies, for example, it would be pointless to try to separate one institutional cluster completely from the other. There are affinities between them that make it possible to "speak of 'industrial capitalism' as a type of production order and as a form of society" (145).

For Giddens, people in modernity live in a created environment, and this created environment is the result of the changes modernity (capitalism and industrialism) has brought. The relationship Giddens establishes between capitalism, industrialism, and environmental change is found in several of his works, but some of his first reflections on all of these variables appear in *The Nation-State and Violence*. At the same time, they are presented in this work in a way that anticipates his analyses of the risk profile of modernity. For example, Giddens notes that the coincidence of capitalism with industrialism, as was the case in European countries, "is the initiation of a massively important series of alterations in the relation between human beings and the natural world" (1985, 146). At the same time, he points out at this point that these changes brought about by capitalism and industrialism are mediated by the development of urbanism.

While Giddens acknowledges that the environment has changed throughout history due to the presence of humans on the planet, the environmental changes in modern times are much more significant in terms of intensity and magnitude. Even in the class-divided societies of the past, where large-scale irrigation systems prevailed, production did not significantly alter nature (Giddens 1985, 146). At the same time, Giddens suggests that in past societies, because people lived in rural areas, there may have been some symbiotic relationship between human communities and the natural world. In many parts of the world in the premodern era, people lived "in" and "with" nature, where it was possible to have a symbiotic relationship between humans and nature. However, he notes that the "advent of industrial capitalism alters all this" (146). Modernity, then, brings with it a significant disruption of man's relationship with the environment.

According to Giddens, the most important change that alters humankind's relationship to the environment in modern times was brought about by the advent of the created environment, a term he will use extensively in both this and later works to capture the environmental changes brought about by the modern era. The created environment finds its most important expression in

the modern urban environment, but it is not something that can be reduced to
that. Nor is the term synonymous with the concept of the "built environment,"
for reasons we will briefly explore below. For Giddens, the created environment in modernity becomes the milieu of all social action, and in it "the
transformation of nature is expressed as modified time-space" (1985, 193).

In both *The Nation-State and Violence* and later works, Giddens holds to
this view. At the same time, this view is accompanied by other statements
that do not always seem to agree with it. For while Giddens takes a multidimensional view of the environmental changes caused by modernity in the
arguments just considered, he approaches the same issues from a narrower
perspective in other parts of his work. In *The Nation-State and Violence*, for
example, where he makes the arguments we have just considered, he states
that the energetic dynamism produced by capitalism have brought about a
transformation of the natural world in a completely different way than in the
past, but that these changes are "intrinsically linked to industrialism rather
than to capitalism as such" (1985, 312). In another place in the same book, he
states that the changes in the natural world are "inseparable from industrialism and not from capitalism as such" (313). This one-dimensional view is not
limited to his statements in *The Nation-State and Violence*, for we find it in
later works as well; for example, in *The Consequences of Modernity* he states
that industrialism is the most important dimension of modernity through
which man's relationship to nature in modernity is determined (1990, 60).
The argument reappears in *Beyond Left and Right*, where Giddens claims that
industrialism is the most important process of modernity that substantially
changes our relationship to nature (1994a, 100).

Giddens's interpretation of the environmental changes that exist in modernity raises some problems that must be considered here. These statements
present us with some difficulty in understanding his views on modernity and
environmental change. In some passages, this relationship is seen as multidimensional, with the created environment seen as the result of the influences of the various dimensions of modernity (industrialism, capitalism, and
urbanism). At other times, however, this relationship is perceived as more
one-dimensional. In the latter case, the created environment is essentially
seen as the result of the influences of industrialism in modernity. The multidimensional view contained in the former case is apparently replaced by a more
one-dimensional view in which, for Giddens, environmental change appears
to be more closely related to industrialism. The differences noted in these passages have made his arguments a target for criticism. According to Goldblatt
(1996), this shift in argumentation has led Giddens to offer a simpler view
of the environmental changes brought about by modernity. For Goldblatt,
the environmental changes originally associated with the earlier triad of

capitalism, industrialization, and urbanism have been reduced to a single dimension that explains the environmental changes caused by modernity.

A first point we need to consider in order to understand this critique is the indeterminacy of the environmental changes Giddens proposes in these various passages. One reason for this confusion is that Giddens does not describe in detail what environmental changes he is referring to in each case. What changes is he referring to when he directly links environmental change to industrialism or to capitalism? We will see later that Giddens attempts to examine some of these changes using the expression "the end of nature." But again, some authors who examine Giddens's ideas show that they are somewhat evasive on this point too.[11] The problem is that Giddens makes arguments that change over the course of his work, as Goldblatt (1996) shows. Is there a logic in the changes of arguments that Giddens shows in these passages?

Let us look briefly at why Giddens has different arguments about these issues and why they may be less contradictory than they seem. One way to look at the relationship between industrialism and environmental change is that the environmental changes associated with industrialism are related to the dimensions that constitute that concept. The environmental changes are caused by (a) the inanimate use of energy sources (fossil resources), (b) the mechanization of production, and (c) the global expansion of manufacturing industry (Giddens 1985, 136–37). In those areas where we can link the occurrence of environmental hazards to these processes, we can link these changes to industrialism. But this, of course, raises an analytical question: Is fossil resource use a problem related only to industrialism, or is it also an economic problem related to capitalism? Although Giddens does not give a clear answer to this question, it could be approached as follows. The elements that Giddens attributes to industrialism are not specific to capitalist societies but are processes that are also present in countries of "actually existing socialism." The use of fossil resources and the mechanization of production, for example, became the productive axes of both capitalist and noncapitalist countries in the twentieth century; therefore, these features can be considered elements of modern societies, capitalist or not. Thus, the environmental impacts that result from the use of material energy sources occurs in conjunction with market mechanisms. This will happen in countries where the industrial system has developed simultaneously with economic liberalism; however, they may have similar or worse effects if these resources are used without market mechanisms. The type of economic system can change the intensity of environmental impacts caused by these processes associated with industrialism, but cannot completely eliminate them if all other variables (e.g., population) remain the same. However, the most important aspect of this view is that the constitutive elements of industrialism must always be considered in the context of the economic system, because industrialism does not function separately from the economy, but only from it.

Moreover, in examining the relationship between industrialism, capitalism, and the created environment, we should keep in mind that a number of possibilities may arise. If in certain cases it seems important to relate environmental changes to diffuse causes, in other cases it may be appropriate to attribute these changes to more narrowly defined causes. Thus, as we have seen above, we can note that Giddens associates the environmental changes connected to the emergence of the created environment with various modern institutions, while at other times he seems to emphasize the influences of a single institutional dimension of modernity. In short, although in his early works Giddens is concerned with urbanism and the human experience associated with it, in later works he turns to the environmental effects of science and technology intertwined with modern industry, things that he, in turn, tends to associate with industrialism. The shift in Giddens's argument in later works does not mean that he rejects his earlier vision, nor does it mean, for example, that he does not consider the effects of capitalism on the environment, as we will see below.

In addition to these points, it should be kept in mind that industrialism is an institutional dimension of modernity and certain risks it entails must be understood through its own dynamics. For example, in *The Consequences of Modernity*, Giddens recognizes a relative autonomy of industrialism in terms of the impact it has on the created environment in modernity. Although in many cases industrialism is set in motion by the logic of capitalist accumulation, he points out that "once under way [it] has a dynamism of its own" (1990, 169). Giddens thus echoes the view of authors such as Jacques Ellul and other thinkers on technology who assume that technological innovations, once institutionalized, tend to develop their own logic, or, as he writes in *The Consequences of Modernity*, "a strong inertial quality" (169).

While capitalism involves economic relations between economic classes, industrialism presupposes a more direct relationship with external environmental conditions that can become the object of human intervention. The various institutional axes of modernity are thus linked to the various relations of exploitation that they can produce. In this case, industrialism is more closely linked to the relations of human exploitation of the natural world. For Giddens, economic class relations are limited to the institutional dimension of capitalism in modernity. Exploitative relations in the realm of politics relate to government power and the means of violence; on the other hand, industrialism, "another important institutional grouping of modernity, concerns primarily exploitative relations between man and nature rather than social relations as such" (1989, 265).

The fact that Giddens associates certain important environmental changes with industrialism does not mean that he ignores the influence of, for example, capitalism. However, when environmental issues are viewed from the economic perspective of capitalism, their analysis tends to change as well. In *The Consequences of Modernity* and also in later works, Giddens underlines the problematic character of this connection. According to him, there are limits to unlimited capitalist accumulation, in terms of resources, regardless of what technological developments occur in the future (1990, 171). The author's reflections on the possibility of a "post-scarcity order" are directly related to the ecological issue. Giddens believes that in this new economic order, heralded by immanent tendencies in modernity itself, the capitalist pursuit of continuous accumulation will undergo a process of dissolution and the ecological crisis will show us that scarcity is inherent in human life on the planet (1994a, 195). A post-scarcity order becomes necessary when continued economic growth becomes harmful, and it is possible when the ethos of productivism is challenged and the possibility is created for the promotion of other life values (163).

Giddens seems to share the concerns of those who see the process of capitalist accumulation as ecologically problematic. At the same time, his reflections on the post-scarcity system do not preclude modernity from retaining the goal of economic growth. In his interpretation, however, technoscientific development is specifically linked to industrialism and, as a cluster of modernity, has institutional autonomy from the other dimensions of modernity. Therefore, for Giddens (1990), it is not correct to reduce environmental issues to the dimension of industrialism. What does this show us? That depending on the environmental issues discussed, they can be associated with different institutional axes of modernity. While the economic issue linked to environmental problems raises a question of scarcity of resources and the flow of their use, consumption, and renewal, the environmental issues linked to industrialism are related to the risks arising from technological and scientific interventions. It also appears that issues related to the created environment can be associated with different dimensions of modernity (capitalism, industrialism and urbanism), but that for the analysis of certain aspects related to the generated risks, the dimension of industrialism can be more emblematic to capture the erratic character of modernity.

We have so far examined the relationship between industrialism and capitalism and their relationship to existing environmental change in modernity, but we have said little about the actual meaning of the concept of created environment in Giddens's work. Here, we shall explore this issue before proceeding with our analysis.

Giddens's (1985) concept of the created environment—and other concepts such as locality and regionalization—arise from his concern with the dimension of space in social theory. In several of his works, Giddens criticizes the scant attention paid to this issue in social theory. The definition offered by Giddens for a created environment is similar to two other terms used by social scientists. They are milieu and built environment. In *The Nation-State and Violence*, Giddens explains that the created environment can be considered the "milieu of all social action" (193). In urban sociology, the term milieu has a different meaning than "place." The local dimension of the city is constituted by its territorial and physical boundaries, which exist for political, legal, and administrative reasons. The milieu, in turn, is "identifiable by the processes around which the lives of city dwellers revolve" (Jayaram 2010, 50). The notion of the "created environment" as milieu gives the term a special meaning. As Cooper points out, the "environment as milieu is ot something a creature is merely in, but something it has" (1992, 166); that is, it implies a symbolic and cognitive relationship between the human agent and the space with which it interacts, which includes a kind of practical consciousness. The fact that the created environment is a place known to human actors, and known "in a certain way," (Cooper, 1992, 167) is due to another important feature that animates this definition. For the created environment is also socially constructed; it is not a given environment that remains unchanged by human presence. Therefore, the term "created environment" in Giddens has a very similar and, in some respects, the same meaning as the term "built environment" used by social scientists. A brief comparison between these two terms will help us understand the meaning that Giddens (1985) attaches to the term.

The term "built environment" is commonly used in urban studies and refers to human interventions in physical human space. This includes stores, streets, buildings, and other constructions, with the city representing the built environment in its entirety. These things constitute the material environment, which is the scenario in which social relations are structured. The difference between this environment and the natural environment is that humans construct this environment themselves, and all the sociological issues that can be derived from this relationship arise from the process of construction. In this respect, the concept of created environment does not entail any significant changes in meaning compared to the concept of built environment. This is because it refers to the same process, even though it attempts to understand this human intervention in external physical environmental conditions in a different way. In this sense, as Moffatt and Kohler point out, "the built environment can only be defined in contrast to the 'unbuilt' environment or ecosphere" (2008, 249). Therefore, the concept of created environment used by Giddens (1995) is an environment that, like the built environment, is a socially constructed environment and is different from the "unbuilt" environment for this very reason. The unbuilt environment can be seen here as a synonym for the natural environment.

The way Giddens (1995) approaches the concept of created environment suggests that the concept could be understood in the same way as the concept of built environment, at least when the latter is used to refer to the urban spaces in which people live. As we have seen above, he considers the created environment as something different from the cities of the past, not only because the created environment is influenced by processes that did not exist in earlier historical periods (capitalism and industrialism), but also because, as a social phenomenon, it cannot be considered as a physical entity separate from the rest of society. In *The State-Nation and Violence*, he notes that in modernity the old relationship between town and country is being replaced by the sprawling expansion of a manufactured or "created environment" (184). And he also says the same thing in *The Consequences of Modernity*, pointing out that in modernity people live in a created environment that is physical but no longer natural. He adds the observation that this created environment is not "just the built environment of urban areas, but most other landscapes as well become subject to human coordination and control" (1990, 60).

In short, the created environment could not be equated with the built environment if we had to reduce it to the space of cities, since in principle it has no specific spatial boundary. For Giddens (1990), the rural environment represents an environment created in the same way as cities. The concept refers to the transformation of the natural environment into a "created environment" or "socialized nature." In this case, the created environment is much broader than the built environment, at least as it has been used, and also than any other similar concept that confines the environment to a limited space. The concept does not distinguish between rural and urban because it assumes that both spaces, under the conditions of modernity, are permeated by the same influences of the processes associated with modernity (capitalism and industrialism).

In *The Consequences of Modernity*, Giddens (1990) deals more systematically with the effects of the social and environmental changes brought about by modernity. In this book, he attempts to examine how these changes in modernity are linked to the phenomenon of globalization. Despite this multidimensionality of modernity, which makes its understanding very complex, Giddens offers an understanding of globalized modernity through a set of concepts applicable to its different dimensions (industrialism, capitalism, etc.). All institutions of modernity undergo similar processes as they globalize; they must decouple relations from local contexts of social interaction to a scenario of interaction in which time and space are expanded. These institutional axes of modernity thus function through generic processes that can be captured by concepts such as space-time distance, abstract systems, expert systems, disembedding mechanisms, trust, and others. These concepts can be applied to the various dimensions of modernity to understand the process by which they are globalized worldwide. Since environmental change is linked to modernity, we need to examine what some of these concepts mean for Giddens's sociology and how they fit into his interpretation of environmental change in modernity.

The time-space distancing becomes a condition for the disembedding and reembedding of social systems under the conditions of modernity. Giddens uses the term "disembedding" to refer to the removal of relationships from local contexts of interaction and their reorganization in new contexts where actors do not participate in the same physical scenario of interaction (1990, 21). Disembedding mechanisms separate interaction from the specifics of place. Social actors who are distant in time and space can establish and maintain social relations through disembedding mechanisms, even when they are in distant locations. Reembedding, in turn, according to Giddens (79), cannot be seen as the reverse change, but rather as the process of structuring social relations under new conditions of interaction in which the co-presence of social actors does not exist.

Symbolic tokens and expert systems are what Giddens (1990) calls abstract systems. When he uses this term, he is referring to these two concepts. Abstract systems are disembedding mechanisms. They allow social relations to be disembedded from local contexts and reconfigured from new spatial and temporal parameters in which the co-presence of individuals is not possible. Symbolic tokens are "media of interchange which can be 'passed around' without regard to the specific characteristics of individuals or groups that handle them at any particular juncture" (22). The second, expert systems, are "systems of technical accomplishment or professional expertise that organise large areas of the material and social environments in which we live today" (27). All disembedding mechanisms, including symbolic tokens and expert systems, are based on trust, which Giddens (34) defines as the feeling or belief we have that people and systems will tend to behave according to our expectations. This leads us to have certain expectations about certain outcomes and events that should result from the trust relationship. Trust, then, is an expression of the belief we have about our relationship with people and abstract systems (science and technology).

In *The Consequences of Modernity*, the environmental changes caused by modernity are examined using the concept of risk. In this work and in later books by Giddens, this concept is given greater prominence than the concept of the created environment, although the two are related. In any case, from this work on, his reflections on the environmental changes that occur in modernity are increasingly examined from the sociological analysis of risk. This goes so far as to foreground environmental risk in his work when attempting to outline the risk profile of modernity. In works such as that of Leiss and Chociolko, risk is defined as "exposure to the possibility of loss" (1994, 6). For Giddens (1990, 91), risk can be understood in a similar way. For him, risk involves a human decision in which chances and losses must be weighed. What "risk presupposes is precisely danger (not necessarily awareness of danger). A person who risks something courts danger, with danger understood as a threat to the desired outcome" (34). However, a purely individual definition of risk that reduces it to an individual's decision would be inadequate, because in society, the actors who make decisions about risk are institutions and organizations, and, more importantly, many of the risks people face in the modern era arise from being exposed to risks that are intentionally brought about by others.[12]

As we saw in the previous part, Giddens links modernity to the imperative to control the natural world. This tendency to control, he argues, can be found in other historical periods as well. Both in the past, particularly in agrarian societies, and in the present, people have used technological innovations to realize their own interests in transforming nature. "The control of nature was an important endeavor in the pre-modern era," he says, "especially in the larger agrarian states" (1991, 135). This quest for control, however, has become radicalized in modernity. In the last two centuries, which just mark the development of modernity, human intervention in nature has been massively expanded, so this intervention is not limited to a few parts of the globe but, according to Giddens (136), has become globalized. In this process, the ecological consequences associated with the created environment are exacerbated, leading to the end of nature.

The phrase "the end of nature" thus refers to modernity's intensification of control over the material world, in which, as we have seen above, the process of human intervention in nature has been extended so that its consequences are no longer confined to particular places and regions. For Giddens, the ecological crisis has its origins in the dissolution of nature, which he understands as objects or processes that exist independently of human intervention (1994a, 206). The argument from the end of nature reintroduces the themes we saw above about the impact of industrialism on modernity but adds some elements that were not present in earlier work. The emergence of the end of nature refers to the transition from the idea of "nature" to "environment," and in this case, as we indicated above, differs from the various uses Giddens (1991) makes of the idea of created environment, in which changes are associated with particular regions and places within nation-states.[13]

The term "the end of nature" would have little meaning if interpreted literally, for it would be enough to consider the universe to discredit the argument. Therefore, of course, the term cannot be understood in this sense. It instead refers to the fact that processes of nature that functioned independently in the past are being integrated into the human way of life. Climate change can be seen as an example of this. Throughout the history of our planet, climate change has occurred independently of the human way of life. However, to understand climate change in modern times, one must also consider the growing influence of humans on this phenomenon. Much of the process involved in sustaining life on the planet, whether in humans or other species, is influenced in its dynamics by the way humans live under modern conditions. The phrase "the end of nature" tends to be anthropocentric, referring to elements of nature that interact with or are causally related to human lifestyles. Climate change is no longer a "natural" phenomenon but an "environmental" phenomenon if we consider the existing transition in our language between "nature" and "environment" that Giddens refers to.

The expression can be better understood through the concept of the Anthropocene. This concept states that humanity has become a "global geological force on par with other natural processes affecting the Earth system (Lidskog and Waterton 2016, 1).[14] Thus, the Anthropocene thesis makes several assumptions that sum up the concept of the end of nature. For example, Lidskog and Waterton (2016) believe that the "Anthropocene can also be interpreted as an unintended side effect of modernization in a mode of analysis reminiscent of that used by Ulrich Beck and colleagues in the 1990s" (4).[15]

Sociologically, Giddens points out two important implications associated with the expression "the end of nature" In the first sense, human life is detached from nature because our lives take place in humanly created locales (1991, 166). In the second sense, human life undergoes a similar process to the extent that nature "is increasingly subject to human intervention and thereby loses its very character as an extrinsic source of reference" (166). In both cases, human life becomes disconnected from nature because it has been socialized in one way or another. While the first meaning refers to the social effects of modern urbanism, the second is more general and refers to problems associated with human interventions in nature itself. The point is that where this human intervention occurs, we can hardly consider the risks associated with human intervention as external. Whereas the idea of the created environment presupposed the existence of a natural environment, the phrase "the end of nature" tends to undermine this division by regarding it as sociologically unimportant.

The end of nature brings us to perhaps the most fundamental point that emerges in this process. In works such as *The Nation-State and Violence*, Giddens emphasizes the relationship between industrialism and environmental change. In *The Consequences of Modernity* and subsequent works, he reformulates this equation by examining it from the perspective of the relationship between abstract systems and ecological risks. Environmental risks are the result of intervention of sociotechnical systems in nature. Socialized nature is created and managed by technical systems in which science and technology, combined with the power of expertise, exert great influence. This means that nature becomes an internally referential system (Giddens 1991, 144). It becomes an element of administration by humans, and its form and quality depend precisely on this administration. And this human control of nature occurs, according to Giddens, through abstract systems. Abstract systems, such as expert systems, are "systems of technical accomplishment or professional expertise that organise large areas of the material and social environments in which we live today" (1990, 27). In modernity, then, environmental risks emerge from a created environment (or socialized nature) as a result, as he says elsewhere, of the incorporation of human knowledge into our interaction with the material environment (124). These threats that arise from the development of these abstract systems are, in turn, mediated by the impact of industrialism on the material environment (110).

In *Modernity and Self-Identity*, Giddens offers an insightful passage about the nature of the abstract systems he refers to and how these systems affect the everyday lives of people in modernity. The passage is insightful because he gives some very concrete examples to illustrate the issues. In it, he gives us examples of sociotechnical systems that organize the water supply and the sewage system, as well as those that provide us with heating and lighting. These systems, as he outlines, have become a prerequisite for stabilizing people's daily routines in the context of modern life. This is because access to water, light, and waste disposal is readily available to most people in the more affluent regions of the world. As Giddens notes, systems such as these have helped to reduce various uncertainties that have historically shaped people's relationship with the environment. Drinking water, for example, has helped provide a response to the inconstant character of water supply, one of the greatest uncertainties that plagued premodern societies (1991, 135). These systems, he adds, helped provide a regulated framework for activities inside and outside the home.

All these examples offered by Giddens (water, lighting, and sewage treatment) touch on sensitive areas that affect our relationship with environment. All these examples address two important functions that nature performs for humans: (a) providing resources and (b) absorbing waste. Hence, he notes that human control of the material environment in modernity presupposes the existence of abstract systems such as these (1991, 135). Since the quest for control in the modern era is hopelessly technological and scientific, our relationship with nature is mediated by these systems. We could mention here the food system, the transportation system, and many others. In all areas where we use and transform resources, these processes are mediated, according to this view, by abstract systems that integrate scientific and technological knowledge. And in all these areas, the abstract systems are usually essential to our way of life. They create the material conditions that generate the routines that constitute them, but at the same time they create a new set of risks.

An important point to consider in Giddens's (1991) analysis of these abstract systems concerns the balance of risks and rewards they generate. On the one hand, these abstract systems are directly linked to a regular and stable supply of resources to people that did not exist in earlier historical periods. In a sense, they can be seen as a prerequisite for the emergence and development of the modern lifestyle in terms of regular and stable resource use by people. Abstract systems enable a regular and predictable supply of resources and thus the creation of a set of routines that shape the daily lives of thousands of people. And to the extent that they enable the creation and maintenance of daily routines related to environmental resource use, these abstract systems are the generators of the ontological security that underlies our relationship with the material world. Giddens understands ontological security as the expectations we create regarding the predictability of the world. The sense we have that events in the world tend to show themselves in the way we expect and know they will (1991, 243). Abstract systems create this confidence because they provide a response to the unpredictability of nature by providing a regular and predictable base of resources and allowing people to make stable lifestyle choices. These choices can then take the form of routines and habits that constitute people's lifestyles. This human intervention in the material conditions of life through these abstract systems (i.e., the socialization of nature) thus makes it possible to stabilize a whole range of irregular and unpredictable influences on human behavior whose source was nature itself (135). In a sense, given the unpredictability of certain functions of the environment in the premodern era, modern life has become more secure and predictable when viewed from these perspectives.

So we can say that abstract systems have helped to create a world of security and predictability for millions of people and the lifestyles that depend to some degree on this regular supply of resources. Abstract systems thus make it possible to colonize the future. In modernity, according to Giddens (1991, 3), the future is continuously colonized by the reflexive organization of knowledge environments. For to the extent that they allow us to deal with the unpredictability of nature, they provide answers to future situations that would be stressful for people and communities. In terms of ontological security, then, they create a seemingly predictable future. Abstract systems become a support for ontological security because they create a possible environment for "calculable actions" in Weber's sense. As abstract systems create the conditions for the development of routines that become integrated into the conduct of life, they also create the conditions for the support of ontological security itself. For, as Giddens (1991, 167) says, ontological security is maintained primarily through the routines of daily life.

If in premodern traditional communities environmental hazards arise from a nature that humans have not interfered with, in modernity hazards arise from a created environment (or a socialized nature) as a result of using knowledge to control the material environment (Giddens 1990, 124). Thus, if abstract systems are directly linked to the creation of security in the modern world, they also prove to be a source of new risks (125). The functioning of abstract systems is an expression of what Giddens calls reflexivity or internal referentiality. This process refers to the ability of these systems to become autonomous and to organize themselves from their constituent elements (Giddens 1991, 5). Therefore, Giddens also states that environmental threats in modernity not only emerge from these abstract systems but are also the result of the reflexivity that these systems foster. In this case, new risks also emerge from the reflexivity of modernity itself (5).

Giddens offers a short list of ecological risks that can be directly linked to the abstract systems of modernity. These include environmental disasters, radiation from severe accidents at nuclear power plants or from nuclear waste, chemical pollution of the oceans, the greenhouse effect from pollutants in the atmosphere, and the depletion of millions of acres of topsoil from the use of fertilizers. Environmental hazards such as these contribute to various changes in the existing risk profile of modernity. Globalized environmental risks are generally characterized by the intensity of the damage they can cause. The globalization of environmental risks also occurs through the expansion of possible contingencies. Climate change, for example, can affect different parts of the planet in different places and at different times, or even simultaneously. Some types of environmental risks tend to be global because of the globalization of modernity itself. Environmental risks of great magnitude are therefore associated with the globalization of modernity itself. More specifically, with the globalization of industrialism and the effects abstract systems such as sociotechnical systems produce that enable humans to control the social and material environment.

However, this process tends to lead to a paradox. All dimensions of modernity are globalized because, according to Giddens (1990), they can bring about a process of time-space distancing of social relations. This is of obvious importance for understanding the environmental problem under the conditions of modernity. As the dimensions of modernity globalize, so do the environmental risks of these processes that shape modernity. Therefore, ecological globalization, with the emergence of global environmental risks, tends to be a predictable process resulting from these changes. Abstract systems thus lead to a time-space distancing of humans in their relationship with nature. The transformation of nature through the production of goods is increasingly accompanied by a global division of labor. Scarcity loses any local character and can be artificially created by an ultimately global production process. The same process takes place with environmental pollution. At the beginning of industrialization, the effects of pollution tended to be regional or national, but today they can reach a transnational or global scale and become detached from their place of origin. In this new context, people and organizations may, on the one hand, distance themselves (spatially and temporally) from the environmental impacts they cause or, on the other hand, suffer from the very environmental impacts caused by others who are spatially and temporally distant. According to Dickens, this whole process leads to a paradox:

> On the one hand, nature is becoming increasingly socialised. In that sense society and nature are indeed becoming increasingly integrated. But on the other hand it is precisely through such socialisation, and the attendant spreading of the social relations, and institutions involved in its production that people lose tangible association with the processes and mechanisms of the rest of nature ant the circumstances surrounding their manipulation. (1992, 150)

Recently, Giddens, in his book *The Politics of Climate Change*, gave this paradox its own name, calling it the Giddens Paradox. This paradox states that people remain immobile in the face of the dangers of global warming precisely because these problems do not seem tangible or visible to them in their everyday lives. This behavior, in turn, leads to postponement of responses to the problem, making it impossible to respond effectively (2009, 2). The feature of environmental risk tends to be inflexible, which in turn tends to reinforce the risk itself. The same argument, as we have already seen, is found in Beck's work (1992).

Giddens's assessment of the environmental risks associated with abstract systems tends to be convincing, but at the same time it seems to lack something fundamental. It is easy for us to accept his diagnosis because, as he himself notes, the risk profile of modernity is accompanied by a well-distributed awareness of risk. Consequently, the list of risks he offers us is common knowledge and, as he notes, create "a numbing sensation, almost a feeling of boredom" (1990, 127). That he takes as given the relationship between abstract systems and ecological risks seems compelling, for we have heard not only of the existence of these risks but also that they are associated with modern science and technology, with which we are also familiar. After all, the risks associated with nuclear waste can hardly be separated from these things. But what is it about abstract systems that leads to these risks?

One answer that seems implicit in Giddens's work is that abstract systems ultimately fail to deliver on their promises and undermine the trust we place in them. The monetary system of a region or a country can collapse; this can even happen to the global monetary system. Even a sociotechnical system designed to provide a secure foundation for water resources can fail. In *Modernity and Self-Identity*, Giddens (1991, 136) points out that a prolonged drought produced by a centralized water system can generate results even more perverse than periodic water shortages. Moreover, abstract systems may be less reliable than "old nature" under certain circumstances (e.g., climate change). Socialized nature may prove less reliable than old nature because it is impossible to know how the external environment will behave in such cases (137). Abstract systems thus generate new risks, which in turn introduce new unpredictabilities for which the abstract systems were not designed. Moreover, one must consider that in the context of modernity, old nature is disappearing as an option. This seems to suggest that the abstract systems of modernity are less reflexive than Giddens assumes. They pretend that their own failure is unlikely and permanently enjoy a promised security that may ultimately fail. For Giddens, abstract systems therefore produce risks that, when realized, destroy the real and perceived security associated with them. It seems that in such cases the trust we place in technical systems breaks down, and social order—understood in terms of the routinization of social life—threatens to disappear.

The created environment is not only associated with risks that introduce new and worrisome environmental hazards; it also has implications for ontological security. When abstract systems fail, bringing with them myriad environmental risks, our ontological security collapses. Giddens's argument does not say, as one might think, that the evolution of the created environment extinguishes the sources of human ontological security; instead, it shows that the maintenance of ontological security tends to remain fragile because of the constitutive role of abstract systems in social life. As far as existential and moral issues are concerned, the created environment seems to entail more losses than gains, for although it provides security and regularity, it is at the same time psychologically and morally unrewarding (Giddens 1989, 279). This assessment seems to involve a paradox related to the emergence and evolution of the created environment. In some respects, modern societies are materially secure, but at the same time they provide a fragile basis for main-taining ontological security.[16]

Finally, there is the effect on tradition itself, at least where it still exists. What distinguished premodern societies in relation to the environment was the fact that a peasant's relationship to his environment was mediated by the prevailing tradition, which led the premodern worker "into an intimate and cognitive interaction with nature" (Giddens 1995, 153). The change brought about by modernity and the development of abstract systems means that tradition, as a cognitive and moral component in our relationship with nature, tends to weaken or disappear. Increasing scientific and technological interventions in nature are causing traditional ways of knowing and relating to nature to change. The "intimate and cognitive" relationship between the farmer and the land, for example, is disappearing. Thus, Giddens (1994a) argues in his later writings that the end of nature should be analyzed in parallel with the end of tradition. The counterpoint to this movement would be the incorporation of socially constructed technoscientific knowledge to mediate this relationship.

In *The Consequences of Modernity*, Giddens mentions the possibility of humanizing technology on the basis of utopian realism.[17] He writes that concern for environmental degradation is pervasive in every government in the world. For him, it is necessary to pay attention not only to the "external impact, but also the logic of unfettered scientific and technological development will have to be confronted if serious and irreversible harm is to be avoided" (Giddens, 1990, 170) in order to avoid serious and irreversible damage. In this book, Giddens suggests the possibility of a humanization of technology, a process that would involve "the increasing introduction of moral issues into what is now a largely 'instrumental' relationship between people and the created environment" (170).

In this work, Giddens does not go much further than this proposal to design future scenarios based on utopian realism. At the same time, he considers social movements capable of providing "glimpses of possible futures" and acting as "vehicles for their realisation" (Giddens, 1990, 161). Social movements could be seen as a reaction to the consequences of the institutional clusters of modernity. In this logic, the environmental movement would be a reaction to the institutional dimension of industrialism (Giddens 1985). At the same time, in his view, social movements are not the only paths that can lead to a safer world (Giddens 1990, 161). Therefore, in his recent publications, he has returned to a more partisan and ideological debate, focusing on the left and the right and on issues of restructuring the state. Next, we will focus on his reflections on environmental policy, which he sets out in *Beyond Left and Right* and later works.

ENVIRONMENTAL POLITICS IN A RUNAWAY WORLD

In *Beyond Left and Right*, the ecological crisis is placed at the center of the possibility of a political renewal of the left, so much so that Giddens states in this work that "the ecological crisis is at the core of this book" (1994a, 19). This book attempts to revive the current guidelines of radical politics. However, it is not always clear how the ecological question fits into the project of grounding this new radical political agenda. Although Giddens connects the ecological question to several other themes, such as the politics of life, emancipatory politics, welfare reform, and the reflexive project of identity, he never fully develops the connections of these themes in relation to the ecological question.

An important aspect of Giddens's critique of the left and the right in *Beyond Left and Right* is that he extends it to environmentalist thought itself. Let us consider his critique first for the first case and then examine the extent to which he believes these issues extend to environmental thought itself. For Giddens (1994a), contemporary ideologies (socialism, neoliberalism) are unable to provide a political response to such environmental risk, as both seem to rely too much on knowledge as an instrument to control change. He suggests that we need a radical approach to address the risks and uncertainties of the current era.[18] But the radicalism offered by the left and the right would be riddled with contradictions. For Giddens, it would be a mistake to associate the left with radicalism, because then the left would have become conservative while the right would have become radical: left focusing on "preserving" the structures of the welfare state, while the neoliberals prove to be "radical" by blindly defending all the consequences of capitalism. A hasty reading of this argument suggests a simple inversion between left and right in terms of radicalism. In another interpretation, Giddens seems to be suggesting something else: that the problem lies in the fact that both the left and the right currently present elements of "radicalism" and "conservatism" simultaneously. And here seems to be the source of today's political paradox: Today, the left and the right combine elements of the two poles. Thus, political radicalism is not exclusively associated with one of the poles of the political spectrum, having lost its close ties to the left.

Giddens believes that these tensions, which affect both the left and the right, also occur in modern environmentalism itself. As he notes in *Beyond Left and Right*, the tendency to see the green movement as an inheritor of the left serves to obscure the movement's affinities with conservative thought; in both cases, the emphasis is preservation, restoration, and repair" (1994a, 11). Thus, advocacy of conservation, especially in conjunction with traditional ways of life, indicates a conservative attitude, while proposals to promote conservation through drastic or even revolutionary change express a radical attitude. Environmentalism thus combines elements of radicalism with others of conservatism, as we have also described for the left and right. Giddens then lists commonalities between conservatism and the environmental movement. Ideologies, then, seem to be becoming increasingly hybrid, uniting values and theses that until yesterday were considered opposites. Even the concept of SD could be seen as an expression of this tension. This concept, like others in environmental thought, resonates, according to Giddens, with the basic strands of philosophical conservatism (201).

This statement should surprise those who regard environmentalists simply as inheritors of the left. For Giddens (1994a, 11), we cannot defend nature in a natural way, just as we cannot defend tradition in a traditional way. What is meant by this? If the environmental risks of modernity result from increasing human interference, attempting to guide environmental policy through the idea of wilderness would be more than a mistake; it would be a way of rendering environmental policy impotent. This is because the idea of a natural environment distracts us from the source of our problems: human intervention in the world. Giddens points out that much of contemporary environmental thinking reflects this view of naturalness. Against this background, the value theory of Goodin (1992), an important representative of contemporary environmental thought, comes under criticism.

Giddens (1994a) offers Goodin's example to show some of the problems with trying to "defend nature in a natural way." According to Giddens, Goodin attempts to base the value of resources on the degree to which they are "natural." In Goodin's view, natural resources are valuable because they "result from natural processes rather than from human activities" (Giddens 1994a, 205).[19] This assumption obscures the challenges facing environmental policy because, as Giddens writes in *Beyond Left and Right*, all ecological debates relate to socialized nature (210). Where these debates take place, we are dealing not with natural systems untouched by humans but with ecosocial systems that involve a socially organized environment. And in these situations, not only would it be impossible to distinguish between what is natural and what is not, but such efforts would also be useless in guiding public policy. For Giddens, recognizing these facts does not mean that nature has become entirely artificial. Rather, it means that no matter what one does, there is no way back to a situation in which this intervention disappears. Thus, for Giddens, any restoration policy implies the establishment of control parameters.

But how, then, should environmental policy proceed? First of all, environmental policy should abandon the emphasis on the "natural." Any environmental policy that aims to preserve nature presupposes some kind of human intervention, which makes it impossible to consider anything as purely natural. This means, then, that environmental policy assumes the end of nature and presupposes the existence of a "plastic nature" (Giddens 1994a, 102). This means that we must acknowledge that in making environmental decisions we cannot refer to that which exists independently of humans (102), because when nature has been socialized, the independence we project into the idea of the natural has disappeared. These choices are not directed toward a world separate from humans but lead us to important decisions about our own way of life. They raise the question: How should we live? (212). For Giddens, then, the most important point of environmental policy is not pristine nature but the processes that expand human intervention in the environment. Just as important as turning policy toward the natural environment, the ecological question leads us to the management of science and technology in a context where both affect the configuration of the industrial system (212). As we have already seen, Beck places science and technology at the center of the politics of RS. Giddens's ecological politics comes back to the same point.

Beyond Right and Left describes some of the uncertainties associated with modern technologies, such as growth hormones and biotechnology. In many of these cases, it is not science itself that is under scrutiny, but the intertwining of science and technology with the modern orientation toward control. Giddens emphasizes that scientific innovation has historically remained within a circumscribed domain (1994a, 215). The "experimental" character associated with new modern technologies, Beck (1992) warns, leaves us living in a "laboratory society." Following this line, Giddens points out that modernity itself has become experimental (1994a, 215). On the one hand, he points out that scientific innovations have an ever-increasing potential to affect our daily lives, and at the same time they are becoming increasingly problematic. Scientific discoveries and the use of technology are being questioned, increasingly challenging the profile of science's impartiality. The components of more "traditional" science (i.e., today's standard view of science) are dissolving. Among these components is scientific impartiality, which requires that scientists not "hav[e] to account for the social consequences of their findings" (217). In many cases involving high-consequence risk, it is not possible to make decisions based on accurate predictions because the consequences of technological innovations are often uncertain and unpredictable. Accusations of alarmism on the part of those who wish to minimize risk or of recklessness toward authorities will remain latent in conflicts over risk. While Beck (1992) suggests that social actors appear to act strategically in pursuit of certain interests that are thought to be involved in the perception of cause-and-effect relationships of risk, Giddens (1998a) sees alarmism and dissimulation as aspects endemic to conflicts over risk. They are not the result of interests alone, but rather the result of the imperfect knowledge that guides social actors' perceptions. As he tells us, "We just cannot know beforehand when we are actually 'scaremongering' and when we are not" (212).

SUSTAINABILITY, MODERNIZATION, AND RISK

The works of Giddens and Beck address several important questions about the idea of sustainability. They provide an assessment of modernity that attempts to describe its condition in terms of risk and relate it to science and technology. Therefore, the work of these authors contains elements that are central to the concepts of SD and EM. Modern risk, especially ecological risk, is a striking feature of modernity. At first glance, sustainability and risk seem to be at odds with each other. After all, how can we think about a sustainable world that is permeated with uncertainties and risks? As Beck notes, "absolute security is denied to us, human beings" (1995b, 19). At the same time, for Giddens, one could imagine a more ecologically sustainable world "beyond left and right." This is a possibility with which Beck (1997a) also seems to sympathize. It now remains to consider how these arguments relate to the concepts of EM and SD.

Beck makes little direct reference to the concepts of EM and SD in his work. In *What Is Globalization?* Beck takes up some views expressed in the discourse of SD. In this book, he acknowledges that it was the Brundtland Report that first recognized that environmental degradation was related to both economic growth and poverty. In this book, Beck (2000, 38) recognizes, as did the Brundtland Report, that these issues should be examined in an integrated manner.

In many passages, Beck criticizes the polluter pays principle, which is considered the central storyline of EM. If the link between causes and effects in assessing ecological risks is imprecise, and if RS produces what he calls organized irresponsibility, what is the validity of this principle? In *The Politics of Climate Change*, Giddens critiques this principle in a slightly different way. He argues that in many cases responsibility can hardly be established in situations involving environmental impacts (2009, 67). In certain cases involving environmental risks, this responsibility cannot be operationalized, as is the case, in his view, with the effects of climate change. On the other hand, he continues in this book, the principle may have important and practical implications for climate policy if some of these obstacles can be overcome. In this case, the literal interpretation of the principle, although it may seem difficult to apply, could be useful in some situations to distribute the costs and benefits of environmental policy.

Giddens mentions the concepts of SD and EM in his works *The Third Way* (1998b) and *The Third Way and Its Critics* (2000). He points out that excessive optimism about market solutions to ecological problems is dangerous, and recognizing that this is the case makes addressing these two concepts important for environmental policy (Giddens 1998, 55). Giddens seems to question the efficacy of the appeal to future generations that the concept of SD entails, and furthermore concedes, as do many others, that the concept is imprecise. In *The Politics of Climate Change*, Giddens approaches the concept of SD "more of a slogan than an anlytical concept" (2009, 63). In this part of the work, Giddens borrows from the observations of Lafferty and Meadwcroft (2000a), who, in *Implementing Sustainable Development*, remind us of the various lists of goals associated with the concept of SD. In this text, the authors point out that SD includes meeting basic needs, protecting the environment, considering the well-being of future generations, promoting greater equity between rich and poor, and, finally, promoting citizen participation in decision making. Given these goals, Giddens notes that such a list empties the concept of any meaning (2009, 62). He makes the same argument in *The Third Way*, where he notes that rather than presenting us with a precise formula, SD is presented as a guiding principle (1998b, 56). SD is a political slogan for Giddens (2009), but in seeing the concept of sustainability as a useful concept, Giddens himself seems to proceed like many other works in the literature. He believes that decision-making processes should be guided by new indicators of sustainability, while suggesting that we should rethink the idea of development.

Giddens's argument is problematic in several respects. First, his proposal is no different from other work that, in using the concept of sustainability, suggests that decision-making processes should be guided by indicators of environmental quality and that economic growth itself should be modified. Second, he seems to assume that the political disputes over the term would disappear if we set it aside, or he assumes that these disputes make it impossible to give the term any important meaning. The fact is that rejecting the concept of SD does not eliminate existing disputes over environmental policy. It is reasonable to imagine that the concept is a reflection of the existing political disputes and not their cause. In fact, in the absence of a common vocabulary, the problem is likely to be exacerbated, especially if the proposal, like so many others, attempts to reconcile economic goals with environmental goals. Third, while it is true that SD can have many meanings when we consider the global debate surrounding around it, this is not true when the concept is examined in terms of the implementation process. This is because in such situations it is assigned a more specific and consensual meaning.

Giddens's view, therefore, runs the risk of ignoring many important changes that the concept entails. It ignores the impact that contestable concepts like this have when they enter public discourse and begin to shape environmental public policy. Jacobs shows that contestable concepts such as SD have two levels of meaning. At the first level, contestable concepts are defined by "core ideas," and at this level we find no disputes about the meaning of the concept, but rather a broad consensus (1999a, 25). In recent decades, according to Jacobs, a consensus has emerged around these core ideas, and wherever they are advocated or implemented, there is little disagreement about their role in reorienting development toward a sustainable course. The core ideas of the concept of SD, to which Jacobs (26) refers, include: (1) integration of environment and economy, (2) futurity, (3) environmental protection, (4) equity, (5) quality of life, and (6) participation. Thus, if you set aside the various existing concepts of SD and look at the core ideas that these definitions espouse, you will find that a significant consensus has developed around them in recent decades.

Jacobs develops a number of observations about the implications of these core ideas of the SD discourse, which will be briefly examined here. First, we should keep in mind that definitions of SD usually include one or more of these dimensions, while it is more difficult to find definitions that include all or most of these core ideas. In any case, according to Jacobs (1999a), each core idea represents substantial value. While it is true that SD policy rarely integrates all of these dimensions, when one of them is incorporated into environmental policy, it can lead to new configurations of development itself. Jacobs points out that in considering these core ideas, we cannot, of course, say that the global economy has followed these principles. However, several of these core ideas have found their way into the rhetoric and actions of governments around the world, suggesting a shift that should not be underestimated. Importantly, once governments commit to one or more of these core ideas, important policy changes can emerge in the process. Given this commitment to the principles of SD, many governments have publicly embraced SD discourse. It is not uncommon for these changes to be accompanied by a process of making the political arena more open and opening channels for stakeholders to participate. As Jacobs points out, "requirement to 'do something' has infected local government and large businesses as well" (28). As a result, Jacobs argues, the commitment to discourse SD has given interest groups and the media a powerful weapon to expose the gap between discourse and action in different parts of the world. Finally, Jacobs points out that in the context of these policy changes triggered by the SD discourse, it is possible to examine the existence of institutional learning in many places. In different places around the world, "the sustainable development discourse is pushing institutions to reappraise their policies and policy-making processes" (29).

We have taken the time to present these arguments of Jacobs (1999a) about the contestability of the concept of SD for a very simple reason: If Jacobs is right about the changes that are currently taking place in the incorporation of SD discourse into public policy, then the thesis that SD is merely a slogan or a vague idea may pose a problem for us when we actually see the changes that eventually take place around the concept. Thus, the worldwide disputes over the term do not prevent some definitions from proving coherent and robust in certain cases. And, more importantly, they give rise to substantive environmental policies that put a new face on development. So the problem for those who see the concept as "vague" is that they are suggesting that nothing is happening in public environmental policy in today's world. And when they realize that this is not true and that something important is happening, then the criticism seems to lose its justification.

Aside from the direct references Giddens makes to the concept of EM, one can see that some of the presuppositions that exist in his thought converge with this discourse. We have seen that the theorists of EM emphasize those processes of social change that they believe make it possible to decouple economic growth from the growing demand for energy and natural resources. Although Giddens emphasizes the importance of these processes in his later works, his reflections on the post-scarcity order tend to emphasize a transformation of the economy in which one of the most important features of capitalism, its pursuit of continuous and spatially extended accumulation, tends to be diluted. These passages seem to indicate that beyond the issues associated with science and technology, Giddens attaches great importance to the changes leading to a new economic configuration of the world economy.

There is also a similarity between Giddens's ideas and those advocated by ecological modernizers. Giddens seems to assume that such a post-scarcity system can include a more ecologically careful economy without necessarily excluding the possibility and feasibility of economic growth—a point of view, incidentally, very similar to that advocated by the EM discourse, as we have seen. Moreover, following an existing assumption in the theory of EM, Giddens argues in his introduction to *The Global Third Way Debate* that the knowledge economy can have great benefits for contemporary environmental policy. According to him, "information technology is intrinsically non-polluting" (2001, 12).

Beyond all of these points, looking at how these approaches examine the relationship between environmental risk, science, and regulation helps us examine their convergences and incongruities more closely.

Giddens notes that SD is linked to and maintains a close relationship with the concept of EM. He notes that SD fits well with EM (Giddens, 1998b, 57). He endorses EM because it manages to combine elements of interest to the environmental movement with others of interest to social democracy. However, if he considers SD a vague concept, EM fails on issues of science and risk. For him, EM tends to sidestep what he sees as essential to environmental policy: the progress of science and technology and their impact on the environment. For Giddens, the discourse and policy of EM sidesteps the very issues that have to do with science and the factors that influence our response to environmental risk (Giddens, 1998, 58). Although EM has some positive points, such as its intention to provide guidelines for environmental restructuring of the economy, it fails to problematize the relationship between science and sustainability. Giddens introduces this critique by examining the postulates of the EM discourse in *The Third Way*. His assessment suggests that EM, as presented, is steeped in a technocratism that overshadows the moral issues of environmental policy. The critique that social scientists generally make of the theory of postindustrial society by turning sociology into an ideological artifact of that society reappears in his work.

Some authors claim that the concepts of SD and EM are based on narrow scientism. Giddens also criticizes both concepts for the same reasons but acknowledges that the literature dealing with these concepts is not completely blind to problems related to science and sustainability. He notes that while the SD and EM concepts appeal to science and technology, they also propose the precautionary principle as a means of addressing the problematic nature of science and ecological risk. In the literature on EM, the precautionary principle is usually presented as a means to prevent ecological threats.

The precautionary principle is a central storyline of EM (Hajer 1995, 27; Weale 1992, 78–79). However, there is no uniform interpretation of the principle, much less a consensus on how it should be applied (O'Riordan and Jordan 1995, 21). In the context of German environmental policy, at least eleven different meanings of the term are recognized (Weale 1992, 79). The precautionary principle is used in a variety of ways. One refers to scientific uncertainties that guide government action in situations where the ability to obtain accurate information and knowledge about environmental conditions is undermined for one reason or another. The precautionary principle, then, says that policy action should not be, and need not be, bound by absolute scientific certainty. Let us see how O'Riordan and Cameron interpret this principle:

> At the core of the precautionary principle is the intuitively simple idea that decision makers should act in advance of scientific certainty to protect the environment (and with it the well-being interests of future generations) from incurring harm. It demands that humans take care for themselves, their descendants and for the life-preserving processes that nurture their existence. In essence, it requires that risk avoidance becomes an established decision norm where there is reasonable uncertainty regarding possible environmental damage or social deprivation arising out of a proposed course of action. (1994, 194)

For Hunt (1994), the adoption of the precautionary principle reflects a concern with the identification and management of scientific uncertainty. Implicit in most interpretations of this principle is the awareness that scientific knowledge cannot accurately predict the environmental consequences of human activities.

The precautionary principle makes it possible to face two types of situations of scientific uncertainty. First, the principle seeks to answer the problem of evidence: What must we know to protect the environment and with what degree of certainty? Second, there is the question of possible policy responses: Based on such uncertainties, what kind of regulatory policy should be triggered? The problem here is not scientific uncertainty per se, but the value judgments that can be made in the face of it. Depending on the political culture of each country, different strategies may emerge to respond to this type of circumstance. Another important point to be considered is the fact that the precautionary principle is used in conjunction with other principles, for example, the shifting the burden of proof for potential polluters. This principle implies shifting the burden of proof onto those who wish to change the status quo.

Shifting the burden of proof is one way of putting the precautionary principle into practice. Other ways of putting it into practice can be the promotion of research and technological innovation, the implementation of compensation schemes, the use of economic incentives (subsidies and taxation), the stipulation of standards with a greater margin of safety for the environment control, and the development of clean technologies (Bohemer-Christiansen, 1994; Bodansky 1994).

As we saw earlier, the idea of sustainability carries with it the assumption that it is possible to establish limits and safety standards concerning the use we make of the environment. Some approaches to the concept are based on assimilative capacity. Such approaches seek to stipulate a level of environmental quality by establishing pollution indices, which implies acquiring, according to Bodansky, "exact scientific information" (1994, 217). However, stipulating these acceptable indices proves, in many cases, to be impractical. The precautionary principle, as noted by MacGarvin, "it is based upon the realisation that it is extremely difficult to determine 'safe' levels of contamination" (1994, 70).

Based on the precautionary principle, common points between the concept of SD, EM, and the works of Giddens and Beck emerge. The precautionary principle shows that sustainability does not need to be equated with a scientific and technological apparatus that provides exact information about the state of the environment. In other words, we do not need to base political action on scientific certainty. As Beck (1992) also warns, if we link political action to a precise state of scientific information, the results can be greater, with more serious environmental risks. Those who subscribe to the precautionary principle know that the acquisition of accurate and consensual scientific information may prove impossible at certain times; however, this does not make sustainability unfeasible, it simply puts it in a new perspective. As Jacobs (1991) argues, uncertainty is an endemic condition to environmental science. But scientific uncertainty does not invalidate it; it actually makes even more necessary the establishment of environmental goals, which are fundamental to guide, at least along general lines, our intervention in the environment.

One of the most important factors affecting the relationship between science and environmental risk is this imprecise basis for decision making, as Beck and Giddens show. Giddens (1994a) speaks of artificial uncertainties and Beck (1999) of risks, uncertainties, and ignorance. How can we be guided by science if it contains more and more uncertainties? If, as we saw in our analysis of the theory of RS, we consider some of Beck's (1992) suggestions for changing the definitional relationships in arguments about environmental risk, we will find that they reflect the same prescriptions found in the literature on SD, at least when the concept is viewed through the lens of the precautionary principle. Beck's prescription is little different from what we might also find in certain interpretations of the precautionary principle.

The affinities between these approaches are evident in the policy prescriptions associated with the adoption of the precautionary principle. For Beck (1992), an alternative to counter the organized irresponsibility that arises with RS is to change the burden of proof in the conflicts of risk definition. According to Beck, to counter the institutional contradictions set in motion by organized irresponsibility, it is necessary to "to shift the burden of proof, so that representatives of industry and science have to justify themselves in public" (1995b, 6). The call for a "reversal of the burden of proof" is an aspect that is also associated with the precautionary principle that appears in the language of SD. If this convergence can be established in principle, some works consider that this convergence is already taking place in European environmental policy.[20] For this reason, Barry argues that the precautionary principle and the policy prescriptions that can be derived from the theory of RS are very close to each other: "While Beck does not directly talk of the precautionary principle, it is clearly consistent with the main thrust of his thesis, and constitutes an important aspect of the relationship between social theory and environmental risks" (1999a, 158). Likewise, Giddens (1994a) alludes to the need for ecological policy to incorporate some type of prudence when dealing with the uncertainties associated with the risks of major consequences. In a sense, if we consider the concern with normative issues underlying disputes about environmental risks as well as concerns about poverty and globalization, it would be possible to say that RS theory is closer to the concept of SD than to that of EM. By postulating a clear shift in environmental reform following certain predictable steps, EM theory ends up offering an unreflective view of the condition of science and technology under the conditions of a modernity marked by the conflicts described in RS theory. Furthermore, both Giddens and Beck recognize that conflicts over environmental risks enunciate moral conflicts that cannot be addressed through a simple environmental efficiency agenda. As Barry notes,

> The precautionary principle fits with Beck's view that environmental risks and dilemmas are not simply technical "problems" to be "solved" by scientific and/ or economic applications. Rather, in "risk society" the precautionary principle acknowledges the normative character of environmental risks—that is, they are

moral questions about right and wrong and not simply about costs and benefits or technical "problems" and "solutions." (1999a, 160)

The normative dimension also supports these perspectives regarding the relationship between precaution, science, and environmental risk. For Beck, the proliferation of risks in RS and organized irresponsibility are about fundamental civil rights. In his view, the political and sociological issues surrounding environmental problems relate to the "legalized and systematic violation of fundamental civil rights-the right of the citizen to life and freedom from bodily harm" (1995b, 8). Giddens holds the same view, stating that environmental policy would encompass universal values associated with the sanctity of human life and universal human rights (1994a, 21).

Thus, the issues that Beck (1992) and Giddens (1998a) suggest play a role in RS are very similar to those of the concept of sustainability, which we evaluated in the previous chapter, and the precautionary principle, which we just considered. Attfield approaches the precautionary principle on the basis of the same values, claiming that precaution implies basic principles of justice and well-being, including some fundamental rights (1994, 154). The precautionary principle not only aims to address the failure of scientific knowledge and its limitations in guiding our decisions but is also necessary for reasons of justice and human welfare. It is these values that trigger precaution (EM, SD) or prudence (Giddens).

Like the theory of RS, the concept of SD presents the environmental issue from a distributive environmental perspective. Beck (1992) does likewise when he postulates that the emergence of RS is due to distributive conflicts in the environment, and the concept of SD does the same when it proposes that our relationship with the environment does not provide an equitable distribution of environmental goods and risks when the needs of present and future generations are considered. The difference is that DS presents the distributional conflict in the environment in normative terms, whereas RS theory presents it in sociological terms. These conflicts constitute a force that triggers various changes in RS. In any case, even Beck relies on this normativity of sustainability by imagining an environmental conflict projected onto present and future generations.

Considering the closeness these approaches express in relation to the precautionary principle, we can see that the problematization of the role of science in environmental policy is common to both. Thus, it is absolutely not the connection these approaches maintain with scientific rationality, nor their distrust of it, that separates or distances them from each other, but the way they position themselves in the face of the limits and possibilities of regulating science and technology. The concept of SD and certain approaches of EM use the precautionary principle as a way of dealing with scientific uncertainty. For Giddens and Beck, however, precautionary ecological policy alone is not sufficient to deal with the complex relationship between science and ecological risk.

The precautionary principle seems to have several weaknesses, at least when interpreted in a particular way. Some interpretations of the principle tend to view uncertainty as a "provisional state" of knowledge and may therefore lead to excessive optimism about the possibility of obtaining accurate knowledge about environmental conditions. Uncertainty in this context can lead to a preference for creating "more knowledge" in order to make policy decisions. Authors like Wynne (1992) criticize certain interpretations of the precautionary principle that contradict the idea generally associated with it, which suggests that political action should not be guided by the search for exact scientific knowledge. Hunt draws attention to the various interpretations that can be assigned to the concept of uncertainty. According to him,

> Discourses about uncertainty, however, expose the awkward nature of the term. It is used to denote different aspects of knowledge and to imply different responses; it is also used as the generic catch-all signifying all cases where scientific knowledge is disputed, or where scientists themselves consider available evidence to be inconclusive. Confusion therefore arises as to the precise meaning of "uncertainty." (1994, 117)

Wynne (1994) outlines four ways of looking at an uncertainty: (a) risk, (b) uncertainty, (c) ignorance, and (d) indeterminacy. There are different responses for each of these. Wynne points out that the more common discussion of the precautionary principle does not include the indeterminacy approach. In this way, not only scientists but also environmental groups may come into conflict over the "correct" or "uncertain" nature of knowledge. This means that sociocultural factors can influence the interpretation and response outlined in the face of "uncertainty." Hunt argues that "it is the social location of knowledge, far more than science itself, that determines the uncertainty of claims. . . . The point here is that scientific certainty and uncertainty vary from one social and political context to another" (1994, 120).

The precautionary principle does not change the burden of proof significantly, as it requires some kind of evidence in order for the burden of proof to shift to the alleged polluters. It should also be kept in mind that if the precautionary principle is used to prevent dangerous and irreversible changes, consensus still needs to be reached on its scope. After all, should this principle cover only human welfare? Perhaps it is because of these and other problems associated with the precautionary principle that Giddens (1998b, 61) does not view it as a satisfactory solution to controversies associated with modern science and technology. As an alternative to the precautionary principle, both

Giddens and Beck point to the need to think about collective responsibility in order to confront the dilemmas of science and risk. This involves not only the question of how we interpret scientific uncertainty and how we will respond to it, but also the question of who will be responsible for the damage and danger caused. In Beck (1992), organized irresponsibility seems to presuppose as its counterpart the idea of collective responsibility, although the author treats this issue only superficially. The dangers that plague RS "take the form of a 'responsibility crisis' (Ewald) in the system, forcing a rethinking of the problem of attribution and regulation in society" (Beck 1995a, 109). Likewise, for Giddens (1994, 29), the moral dilemmas that the newest risks pose to us imply the elaboration of an ethics of collective and individual responsibility. In his most recent works, Beck has insisted that risk and responsibility are connected. Giddens likewise refers to these themes in his later works. An effective approach to addressing the issue would be to make those responsible for innovations, according to Giddens (2000, 139), take responsibility to a greater degree for what they do.

Strydom (2002) rightly notes that the concept of collective responsibility is underdeveloped in the work of Beck and Giddens. However, this does not detract from the evidence and the few directions to which they have pointed. After all, this is a topic to which sociology as a whole has paid little attention.[21] For authors such as Strydom, the idea of collective responsibility does not mean the absolute prohibition of research associated with hazards and risks, but the creation of new regulations for the critical debate and subsequent regulation of technoscientific development. It is about conducting scientific and technological research more carefully, subjecting its development to the scrutiny of public opinion, while respecting regulation by global institutions.

Despite the controversies surrounding the precautionary principle and the notion of collective responsibility, both represent important starting points for confronting the dilemmas that arise from the complex relationships between science, technology, and ecological risk. In some ways, collective responsibility represents an extension and deepening of the issues contained in the precautionary principle. Thus, these options need not be mutually exclusive, again suggesting that the work of Giddens and Beck does not go in the opposite direction of the concepts of sustainability and EM.

As Boswell (2009, 94) notes, in Giddens, the policies associated with new technological and scientific risks can no longer be expressed in terms of material interventions. This differs, for example, from the policies that emerge from EM, where an integration of sustainability and economic growth is sought. This "win-win" game presupposes a "materialist politics" in which economic processes tend to ensure ecological security in the use of the resource base. However, in areas where manufactured insecurities prevail, reconciliation between the environment and the economy can hardly be established with certainty, not only because the prevailing uncertainties prevent a reliable realization that this integration can be easily achieved, but also because in many cases it is not, or not so much, about economic issues. As for Beck, the issue is the use of scientific knowledge in decision-making processes, considering that this knowledge can be in the form of uncertainty and ignorance.

Unlike Beck, however, Giddens's position is less clearly definable, as his view seems to change in some respects. Whereas in works such as *Beyond Right and Left* he argues for a policy oriented toward prudence, in *The Politics of Climate Change* he proposes replacing the precautionary principle with the percentage principle. This indicates a change in his argument, if we assume that his defense of prudence-led policy in *Beyond Left and Right* is in some sense an endorsement of the precautionary principle itself. At the same time, it shows a divergence of his ideas with those found in SD concepts and also in the theory of RS. As we have seen, the precautionary principle tends to be an important axis of the environmental policies proposed by these concepts, as we have just examined. Giddens (2009) follows the ideas of Cass Sunstein (2005) on this topic and, like the Harvard philosopher, proposes to replace the precautionary principle with the percentage principle.[22]

What is the point of Giddens's adoption of the percent principle and his critique of the precautionary principle?[23] Consistent with this approach, it seems certain that answers to uncertainties cannot come from science itself. If scientific uncertainties are irreducible—meaning that we cannot convert them into certainties by "knowing more"—then our decisions to address the dilemmas posed by high-consequence risks must rest on new foundations. The precautionary principle offers one way to do this by providing a basis for decision making in situations where these uncertainties exist. Giddens (2009), like Sunstein (2005), argues that the precautionary principle is rather illusory because it tends to create several misconceptions. For Giddens, the principle tends to be too conservative because it focuses on one side of risk and emphasizes costs rather than benefits (2009, 57). Thus, the principle is a conservative policy position that then expresses a tendency toward inaction. This is even more evident in the stronger versions of the precautionary principle, which, if applied, would paralyze any kind of innovation. If this conservative tendency does not appear in the weaker versions of the principle, it is confusing because the prescriptions that arise from these versions tend to justify completely opposite courses of action. Thus, if the principle has a clear alignment, it tempts us to be overly conservative in the face of risk. If these orientations are more flexible, the prescriptions to which the principle tends are as diverse as possible. The principle's conservative orientation is incoherent because inaction as a means of avoiding certain risks leads us to neglect the very risks that result from inaction itself.

With these arguments, the replacement of the precautionary principle with the percentage principle seems to be justified. The rationality of the precautionary principle becomes incoherent because, in an effort to avoid certain harmful effects associated with certain risks, we create others that may be just as harmful as those we sought to avoid. The percentage principle provides greater rationality because, by weighing the benefits and losses associated with risks, we are better prepared to make a wise decision. The principle allows us to consider opportunities that tend to be underestimated under the precautionary principle.

Giddens's (2009) answer to these questions expresses two kinds of problems. First, he gives an answer that seems somewhat simplistic. This is because Sunstein's (2005) approach seems to incorporate several of the problems that guide standard risk analyses, which propose a simple calculation of costs and benefits, in decision making. The fact is that existing scientific uncertainties are rather problematic for this type of calculation. Second, in addition to underestimating what the percentage principle can offer, it also simplifies the importance of the precautionary principle. Like Sunstein, Giddens relies on a reading of the existing strong and weak versions of the principle without considering the possibility of an intermediate versions where innovation can be incorporated.[24]

CONCLUSION

In his recent works, Giddens (1998b, 2000) has spoken positively about the role that EM could play in linking (renewed) social democracy and ecological issues, a link that seems to have been crucial in recent years in their attempt to create a Third Way for politics. Giddens has proven to be receptive to the processes of change that have taken place under the discourse of EM. This approach becomes even clearer when we consider his thoughts on the concept of SD, perhaps because, in contrast to this latter concept, Giddens views EM as a minimally coherent program to promote some major ecological changes in the world's industrial sectors. But this seems to suggest that he has more confidence in what EM can offer as a political program than in its status as a sociology, not least because, as he makes clear in his work (Giddens, 1994b), trying to describe current changes through a somewhat linear and predictable lens tends to be insufficient to capture the implications of what institutional reflexivity means.

Giddens (1990) is particularly important in this debate because his own work seems to somehow connect these concepts. His sociological analysis

makes clear the relevance of RS theory for understanding the ecological conditions of modernity. At the same time, his work is sensitive to the positive
role that EM and SD concepts can play in this context. However, Giddens's
view of EM is closer to the work of authors such as Hajer (1996) and Christoff
(2010) when these authors point to the possibilities of EM becoming more
reflexive. Unlike other authors, Hajer (1996) sees EM as a kind of cultural
politics. And unlike other proponents of EM, he offers a more critical reading
of the concept, as Buttel (1997) points out. Regardless of what this interpretation of EM may mean, it is important to note that this perspective does not
view Beck's RS as incompatible with EM, but rather as a perspective that
he believes can provide a critical assessment of EM. Through this critique,
which is furthered by RS theory, the meaning of EM could change in the present and the future. This seems to be consistent with the position of Giddens
(1998a), who views Beck's RS as a foundational work for understanding the
ecological consequences of modernity. In this way, Giddens's work can be
seen as a bridge connecting the views of Beck, EM, and SD. These perspectives, though seemingly at odds with one another, could form an intellectual
and political dialogue that allows for deeper engagement with the challenges
and opportunities of an ecologically damaged modernity.

The concepts of politics do not simply function as lenses through which
we observe a process independent of us, but are part of political life itself.
They help to constitute it and make it what it is. This implies that changes
in these concepts, once accepted by a significant number of social actors,
end up contributing to changes in political life itself (Connolly, 1983). Thus,
concepts such as SD and EM not only provide a lens through which we can
observe the socioenvironmental reality but also exert an influence on the very
construction of the reality they help to analyze. Therefore, Mol has stressed
that it is not possible to separate sociology from the ideology (or discourse)
of EM. In his words, "But I acknowledged at the same time that it is truly
difficult to maintain this formal distinction in 'practice': in the contributions
of (environmental) sociologists to the development and reform of modernity"
(1995, 397). For Mol, the reflexivity of concepts such as these involves a
political dimension that becomes fundamental to what social scientists seek
to prescribe and describe. The extent to which an industrial society is restruc-
tured according to the principles of EM depends on the tensions and conflicts
in the society that are to change the course of modernization. The more the
ideology of modernization is accepted and adopted by the important actors
in industrial society, the more industrial restructuring will take place in the
direction set by EM. But this path, he reminds us, can be reshaped by the new
information and knowledge about the social and environmental impacts that
arise in the development of this new path of modernization. Thus, the path of

EM can be revised and modified on the basis of the institutional reflexivity to which it must be subjected.

If the concept of SD and the work of Giddens (1990) and Beck (1992) bring some theoretical contribution to the EM perspective, it is to be expected that this has implications for the political practice that the EM discourse supports. For the theses of EM to be confirmed, it needs to go through a certain social experiment. But, as the works of Giddens (1994a) and Beck (1992) indicate, any type of social experimentalism, whether it has its origin in the natural or social sciences, should be performed under the brand of collective responsibility.

There are also other elements in the works of Giddens (1990) and Beck (1992) that touch on fundamental aspects of the concepts of EM and SD. As Beck argues, with the social explosion of risk, science suffers a quake of legitimacy and trust. In this process, the uncertainties and conflicts that arise challenge the domain of science, as well as the position of the State as the guardian of public safety. Many of the issues raised by Beck (1992) and Giddens (1990) concern the trust and legitimacy of technical and scientific development, attributes that, as Jacobs (1995) indicates, constitute key elements for any sustainability project.

The relationship between interests and scientific knowledge is also worth mentioning. The idea underlying the theory of RS, that economic and political interests interfere in scientific and technical development, is not new in the social sciences; it is very present in what some authors call the political economy view of science. From this point of view, scientific and technical knowledge is recurrently shaped by commercial and political priorities (Yearley, 1988). The contribution of Beck, in this respect, has been to introduce this assertion into the field of discussions about environmental risks. It is for this reason that Beck (1992, 55) suggests that his sociology of risk would also encompass cognitive (knowledge) sociology and a political theory.

In the 1990s, it was recognized that the incorporation of the environmental theme by sociology required a sociology of knowledge. Buttel and Taylor (1994) list some reasons why environmental sociology would need to follow this path. Among them is the fact that when social scientists are faced with environmental issues, they are faced not with absolute facts in themselves but with interpretations and evaluations coming from research institutes and natural scientists. A sociology of knowledge is necessary to unravel the interests and values underlying forms of environmental knowledge.[25] In summary, what Beck states in *Risk Society* (1992) and his other works echoes many of the interests and concerns raised in this area of knowledge.

If EM is silent on important aspects of ecological reform, Beck and Giddens tell us little about the economic reform that a sustainable society requires. In neither do we find any indication of what an ecological economy would be, in the same way it exists in the literature on the concepts of SD and EM. If an ecological economy seems possible in the RS theory, it would be just a result of the self-criticism promoted by the ecological policy that it encourages.[26]

In the specialized literature, there is an attempt to differentiate the concepts of SD and EM, as well as the various versions existing within them, based on the postulation of "strong" and "weak" models. Others mention the terms "radical" and "superficial" to refer to the different models of SD and EM.[27] It is not difficult to note that there is an attempt to find what would be a radical political model for these concepts. In fact, this discussion only reflects a long debate about the difference between the existence of a "shallower" and a more "radical" environmentalism (Dobson 1990). In the same way that some seek to differentiate between environmentalism (more superficial) and ecologism (stronger), others try to differentiate a weak sustainability from a supposedly radical one. Thus, the problems that affect these concepts go through the ones indicated by Giddens (1994a) concerning conservatism and radicalism. The difficulty in making these concepts politically radical is that they have a distinctly conservative profile. After all, the ideal of sustainability is close to the idea of conservation, which is, in turn, related to political conservatism itself. This is also the case for the appeal to the precautionary principle (or prudence) that both EM and SD express. Due to its origin, prudence seems to be a term closer to conservative than radical thought. These observations are also valid for the theory of RS by Beck. As Eder states, "The idea of a Risk Society is a fundamentally conservative one—it implies a distancing from what makes life risky, a return to a world that avoids dangers" (1996a, 217). Perhaps this is not necessarily a problem, since, as Giddens points out, "we should all become conservatives now" (1994a, 49). In this case, concepts such as SD and RS can be seen as similar, given the political profile that characterizes each of them. EM can contribute to the design of a radical ecological policy if it is in tune with the issues raised by the concept of SD and by the works of Giddens and Beck. Giddens (1998b) is an example of this possible compromise, as he endorses the view Beck's (1992) view of RS and, at the same time, does not rule out the role that the EM and SD concepts can play for a radical politics.

NOTES

1. This critique can be found in Sachs (1993); Moser (1995); Roe (1995); Lash, Szerszynski, and Wynne (1996); and Luke (1995).

2. See Blühdorn (2000), Hannigan (1995) and Christoff (2010).

3. On this debate, see Rifkin (1999), Shiva (1995), and Kollek (1995).

4. For an analysis of the relationship between work specialization and scientific specialization in the ecological context, see the work of Dickens (1996).

5. Influenced by social scientist François Ewald, Beck (1995a) sees security as a sociological phenomenon supported by inventions and institutional arrangements that industrial society has created to deal with the dangers it has generated.

6. Beck (1992) is directly influenced by the work of Charles Perrow (1984) in his considerations of "normal accidents."

7. See also the article entitled "Accidents in High-Risk Systems" (Perrow, 1994) in which the author reconsiders his theory of system accidents.

8. Beck (1992) seems to have incorporated the arguments from Perrow (1984) regarding normal accidents. The theory by Perrow is not about the inevitability of catastrophes but of accidents. The circumstance by which these accidents turn into catastrophes is a topic the author has not analyzed in detail in his work. According to him, catastrophes require the unusual confluence of many conditions that are not present in every system accident (3).

9. An example of a sub-politics to which Beck (1999) refers is the global boycott of the oil company Shell in 1995. On that occasion, Greenpeace, together with citizens from around the world and with the support of some ministers of state, succeeded in influencing the practices of that company concerning waste disposal. There are more countries where this type of political power from citizens and the environmental movement, in alliance with other political forces, has obtained regulatory power over the industry. In the Netherlands and the United States, negotiations on environmental safety standards were performed directly between NGOs and companies, without any type of state intervention.

10. In institutional terms, modernity has four axes of change. The institutional axes that define modernity are: (a) surveillance, (b) military power, (c) industrialization, and (d) capitalism. While other authors debate which of these features are predominant in shaping the modern world, Giddens opts for a more comprehensive view. He views modernity as "multidimensional on the level of institutions" (1990, 12). Each of these axes is linked to the respective processes that in turn shape globalization: (a) the nation-state system, (b) the military world order, (c) the international division of labor, and (d) the capitalist world economy. Thus, for Giddens, modernity is "inherently globalizing" (69).

11. For Dickens, for example, Giddens is "never clear . . . what distinctions he is referring to by the term 'nature.'" What exactly is this 'nature' that is destroyed or 'ended'?" (1999, 102). Dickens notes that "the problem, therefore, is that the 'nature' that is supposedly 'terminated' takes many and different forms in different parts of Giddens' work" (103). The analysis in this book is somewhat different. The meaning of "the end of nature" seems reasonably clear to us in Giddens's work, and we will return to it later.

12. This last point is crucial because in disputes over risk, this process becomes the axis that drives these conflicts. Risk assessments, and policies guided by those assessments, determine the distribution of opportunities and costs associated with those decisions. Therefore, according to Giddens (1990), individuals live in an institutionalized risk environment in which decisions about risk are already fixed.

13. For Giddens (1994b, 77), the change in language we have witnessed in recent decades is an expression of these changes. This is because the use of the word "nature" generally refers to that which is created or exists independently of human action. The word "environment," on the other hand, suggests the opposite, referring to nature that has been transfigured by human intervention. We could say, then, that today's widespread use of the term "environment" is due to the fact that the term "nature" is no longer sufficient to describe ongoing environmental changes.

14. This view is important not only for environmental sociology and all social sciences dealing with environmental issues, but also for the natural sciences. If in the first case environmental sociology must incorporate an understanding of the dependence of social life on earth's ecosystems, then in the second case "environmental sciences need to include an understanding of how ecosystem processes are heavily influenced by social processes. Not only environmental issues, but also the environment itself is co-constituted by ecological and social processes" (Lidskog and Waterton 2016, 1).

15. When this expression was used in the 1990s, the "end of nature" thesis was criticized in several works; see, for example, Dickens (1999) and Benton (1999). However, if we consider the growing convergence of social and natural scientists toward the Anthropocene thesis, the "end of nature" metaphor should be considered important and powerful for thinking about environmental sociology in the Anthropocene. Although there are obvious convergences between the two ideas, we will not analyze these similarities in this work. Maldonado (2018) states that the concept of the Anthropocene is not a new idea but, in his opinion, was already expressed in the works of McKibben, Beck, and Giddens. According to him, the concept of the Anthropocene reinforces the metaphor of the "end of nature" presented by Giddens in the 1990s.

16. Giddens's ideas on these questions are reminiscent of Durkheim's own assessment of the social consequences brought about by modernity. Durkheim acknowledges that the modern world allows for the creation of a material abundance that is incomparable to earlier historical periods. On the other hand, this material security by no means provides a guarantee of happiness and human flourishing. In *The Advent of the Division of Labor*, Durkheim presents us with the following doubt: "But is it true that the happiness of men increases in proportion as men progress? Nothing is more doubtful" (1984, 186). According to Durkheim, modern times not only offer no guarantees of greater happiness, but they open up a precondition for various anomic states of the individual, one of which is suicide. Incidentally, both Saunders (1989) and Benton (1999) see reflections on Durkheim's analysis of anomie in Giddens's (1991) analysis of the created environment. Saunders notes that in his analysis, Giddens "goes further to relate Durkheim's concept of 'anomie' to all of this by arguing that the created environment creates an anomic crisis tendency in contemporary societies" (222).

17. Utopian realism, in Giddens, refers to political projects for change inspired by a combination of realism and idealism. This utopian realism would be a legacy left by Marxist thought and is evident in Giddens's when he argues that "we must keep to the Marxian principle that avenues for desired social change will have little

practical impact if they are not connected to institutionally immanent possibilities" (1991, 155).

18. According to Giddens (1994a), radicalism has had two basic meanings in the past: both the desirability of a certain kind of relevant change and the possibility of controlling the direction of that change. At the same time, it emphasizes that radicalism is something generally associated with socialist thought.

19. Giddens's interpretation of Goodin's work seems to contain some problems. He claims that, in Goodin's view, the value of nature lies in those things that are not subject to human influence. But this, as he himself notes, is because for thinkers like Goodin (1992), natural processes are seen as "larger than ourselves." He relies on the following sentence from Goodin's work to support this argument. In *Green Political Theory*, Goodin writes that the "idea that what is especially valuable about the products of natural processes is that they are products of something larger than ourselves" (1992, 38). This, however, denotes Goodin's interpretation of environmentalist thought rather than an expression of his own views. For Goodin, the primary concern is not with the idea that natural processes are "larger than us," but with the idea that they are "things outside of us." This, in turn, has different implications than those drawn by Giddens (1994a) from his interpretation of Goodin's (1992) ideas.

20. In examining the regulatory changes that have been made in food policy in Europe in light of the risk posed by genetically modified organisms (GMOs), among others, Stapleton notes that "the evolution of the precautionary principle reveals the entrenchment of Beck's RS, and how the push for incorporation of the precautionary principle into food safety regulation represents the institutionalization of concern about the distribution of risks" (2016, 518).

21. Some exceptions in this picture are authors such as Apel, Jonas, Melucci, Hegedus, and Delanty. It should be noted that, for Strydom, "an awareness of this normative dimension was clearly signaled by the Brundtland Report" (2002, 128).

22. Sunstein (2005) critiques the precautionary principle can be found in his book *Laws of Fear*. Giddens uses Sunstein's critique to invalidate the use of the precautionary principle in environmental policy.

23. We must bear in mind that Giddens (2009) adds two important observations related to the application of this principle. The percentage principle should work in a democratic way, i.e., the decision about the risks in question should be made in an open framework of public debate. At the same time, risk assessments should take into account the values present in the social context in which the risks are considered.

24. For a critical response to Sunstein's work, see Sachs (2011) and Mandel and Gathii (2006).

25. Buttel and Taylor (1994, 228) recognize that environmental sociologists have little knowledge of modern debates involving studies on scientific knowledge and technology.

26. However, this void in Giddens's work is being filled by authors who sympathize with his political ideas. In *The Environment, Modernity and the Third Way* (2001), Jacobs presents some guidelines for the construction of an ecological economy in the context of debates about the Third Way. The author also presented a summary of his ideas in a short text that was suggestively titled "Environmental Modernization"

(Jacobs, 1999b). Finally, and as we seek to show in this work, the guidelines Jacobs (1991) launches for the construction of a "green economy" are very close to the ideas defended by some ecological modernizers, as we could see in the previous chapter.

27. See, for example, Christoff (2010) and Hajer (1995).

Conclusion

This work was conceived out of our dissatisfaction with a large part of the bibliography on the approaches to EM, SD, and SR. A common view that has prevailed is the supposed incompatibility that exists between these perspectives, sometimes disguised under the argument that one approach is more "radical" or "deeper" than the other or masked under the claim that they glimpse different trajectories of social change. Sometimes it is argued that the divergence is in the relationship between EM and SD; sometimes there is a conflicting reference between EM and SR. If social researchers accept this interpretation, a difficult crossroads will arise before them shortly thereafter. An exclusive choice between these theoretical perspectives will have to direct the theoretical and practical interests of environmental sociology. Which one should we choose and which one should we discard?

If the arguments developed in this study are correct, or if they have some relevance, we should start to suspect the view that establishes a rigid opposition between the approaches analyzed here. The first question is: What makes the difference between these approaches? Secondly, it becomes necessary to know whether, with such a difference, this constitutes a real obstacle to bringing them closer together. Throughout the book, we have looked for answers to these two questions. As we have seen, the theory and discourse of EM have a strong economic dimension, as they try to reconcile economic growth and environmental protection through the defense of a process of ecological economic growth. It is understandable, then, that to skeptics of all kinds and advocates of negative or zero economic growth, EM would seem like a politically indigestible alternative. However, defending these last options can turn out to be as dangerous or more dangerous than the first. No economic growth rate (negative, zero, or positive) tells us what is happening to the environment. Furthermore, ecological modernizers are not the only ones advocating ecological economic growth. In any case, the possibility of reconciling economic growth with environmental protection is a central aspect in defining the discourse and theory of EM. It is the central presupposition of this belief system (Weale) and its main storyline (Hajer/Dryzek). In summary, a crucial question for environmental sociology is to assess how the ecological question can be integrated into modern economics. EM provides a preliminary basis for this.

The concepts of SD and sustainability raise a different issue from the one posed by EM. The concept of SD, as expressed in the Brundtland Report, is just one version among many of the idea of sustainability. Both concepts are accused by social scientists of imprecision and could not, therefore, serve as guidelines for environmental sociology. Our analysis led us to the opposite conclusion. The contribution of the sustainability concept, contrary to the EM concept, lies in the fact that it understands the normative implications of the environmental protection process. The question "What must be sustained?" is primarily a matter of morals rather than a scientific order. And it takes us to the subsequent questions involving justice and democracy. Sustainability runs the risk of becoming an empty and imprecise ideal if not guided by these values. Ultimately, our interest in the environment, and consequently in its protection, is due to an anthropocentric interest. According to the Brundtland Report, this interest is to satisfy basic human needs. However, by including a democratic reform in the sustainability package, other "ecological rationales," which go beyond this anthropocentric interest, may also be considered. The fact that we associate "human needs" with "environmental protection" does not mean that there is no interest in the environment itself, whether for aesthetic, religious, or any other reasons. The deliberative democracy that some authors associate with sustainability presupposes that these different interests can be reconciled.

The normative character of sustainability and its ramifications in political ideals of justice and democracy are, in our view, the fundamental contribution of the concepts of SD and sustainability. The restructuring of the industrial capitalist society cannot, and should not, be just *economic*; it must also be *moral*.

Having considered these points, what differentiates the last two approaches from the works by Beck and Giddens? In general terms, the fact that these authors highlight the ambivalent role of scientific knowledge and technology in the ecological policy. In chapter 4, we saw that science and modern technology play a crucial role in political projects involving the concepts of sustainability and EM. Many approaches to sustainability are based on setting limits, or what some authors call the environmental assimilative capacity approach. It is noticed that the primary concern for sustainability is to know how much we degrade the environment. The second challenge is to figure out the technical means of reversing this situation. Science emerges as an essential informational base to assess our impact on the environment, and technology, in turn, not only contributes with the means to understand the processes of environmental degradation but also provides the tools to reverse them.

Beck and Giddens challenge this optimistic view. Beck is convincing in showing that, from a certain point of view, science and technology are linked to the generation of ecological risks. The first reason for this is the social context in which technoscientific development takes place, prioritizing increased production over risk prevention. Second, there is the indication by Beck that scientific rationality itself, disregarding even any social influence that might exist on it, can prove harmful in the attempt to reach acceptable standards of environmental quality. Assumptions about causality, certainty, the distinction between fact and value, specialization, the separation between theory and practice, among other factors, can make us neglect the existence of dangers or distort our own perception of the problems involved. It is easy to understand why Beck's (1992) thesis regarding the RS is considered antagonistic to the proposals contained in the concepts of SD and EM. While the latter seems to celebrate the promises of technoscientific "enlightenment," Beck casts, in an explicitly postmodern attitude, a veil of distrust over such promises: How can we solve environmental problems with what helped to create them?

With all these differences highlighted, it remains to be seen whether they make the three approaches incompatible with each other. If environmental sociology is challenged to envision new paths for economic growth to ecologize the economy, EM theory has an important contribution to make in this area. Since changes involving technology, industry, and economics are needed today, environmental sociology will take, in part, the form of economic sociology.

This economic bias of EM does not clash with the normative profile of the concept of sustainability unless economic growth is seen as intrinsically unsustainable. However, this is not necessarily the case. The economic growth pursued by EM presupposes a demand for greater *inputs* of information and not for environmental resources. In our view, the error of EM theorists lies in restricting the necessary changes to a two-way process: "economizing the ecology" and "ecologizing the economy." The problem is the economic reductionism that some EM interpretations can bring with them. The restructuring of modern societies toward sustainability cannot be just economic, although the economy plays a central role in this process. Changes in the economic sphere are necessary, but are not enough. Other fundamental questions need to be resolved, for example: What kind of democracy will guarantee this change? What kind of justice and what rights should accompany economic restructuring?

Recognition of the normative nature of sustainability is essential to relativize the economicist bias of EM. Economics has been presented as one of the most "exact" among the social sciences. This movement, in turn, has involved a "depoliticization" of the economic phenomena. Much of the literature on EM perhaps reproduces this bias in contemporary economic thought. In doing so, it hides the normative character implicit in socioenvironmental reform. One way in which environmental sociology avoids excessive economism is precisely to recognize that science cannot separately answer the question "What should be sustained?"—a question, by the way, inherent to the concept of sustainability. The ecological restructuring of modern societies implies a process of valuing the environment that counteracts and even surpasses the type of economic valuation that perhaps still prevails in the environmental policy of EM. In this sense, environmental sociology must be seen as a sociology with a normative content that transcends the narrow appreciation of economic models and, if applicable, of EM itself.

This makes environmental sociology a kind of political sociology, focused on usual issues that exist in the area of politics, such as the state, democracy, citizenship, and human rights. This type of approach is not totally against EM, but only its excesses. The guidelines issued by EM, including the financial instruments to promote environmental protection, may be used, as long as the issue of environmental valuation is not reduced to economic methods and as long as principles and objectives such as justice and democracy are included.

As for justice, the concept of sustainability requires that we do not see the basic resources and services provided by the environment as "naturally" guaranteed, but as something made possible and regulated by economic and political institutions. Ecological conditions can, therefore, be fundamental to the realization of political and social rights, which raises the debate about a new era of "environmental rights." There is also, as we mentioned, a close relationship between sustainability and democracy. Many of the aspects involving sustainability invite public deliberation, but representative democracy, as it exists today, seems to be insufficient to achieve this goal.

If restructuring is not reduced to economic change and includes the normative dimension that the concept of sustainability brings with it, Giddens and Beck show us that it is also necessary to problematize knowledge and modern technology. The economic foundations of EM can thus be complemented by the moral reform implied in the concept of sustainability since both EM and SD offer partial answers (precautionary principle) to the problems posed by these authors.

The interpretation that SD and EM constitute "optimistic" perspectives concerning modern science and technology, while the sociology by Beck is more "pessimistic," is not entirely correct. On the one hand, both SD and EM embrace the precautionary principle, which, given its characteristics, represents a recognition of the limits of science as a secure informational basis for politics. On the other hand, Giddens and Beck admit that scientific knowledge is fundamental for the delimitation of risks, however controversial its use may be.

In a way, these questions lead us to moral reform, implicit in the concept of sustainability, since they lead us, once again, to values related to human rights. The precautionary principle, for example, is embedded in values associated with human well-being and justice. Although the principle itself also contains weaknesses, this does not mean we have to discard it entirely. What Giddens and Beck have to say about the ambivalent condition of science does not contradict the literature that emphasizes the normative character of the concept of sustainability and its association with the project of deliberative democracy. If sustainability engenders a moral dilemma in terms of principle (What to preserve?), the questioning of science in RS poses a new moral dilemma of a cognitive order: To what extent can we trust the knowledge we possess? A certain precaution (EM) or a form of prudence (Giddens) seems to be necessary for ecological policy. In addition, it is also important to anticipate responsibility for possible dangers, regardless of the conflicts that may arise around scientific uncertainties.

With the sociology of RS, we have realized that environmental sociology is also a sociology of knowledge. It needs to incorporate much of the controversies, doubts, and assumptions that RS theory casts about modern science and technology. Although some aspects of the view by Beck on the ambivalent nature of current science can be refuted, the problematization of technical and scientific development is pertinent and inevitable. It becomes necessary to know what we are going to do with the lack of reliable scientific information and with the "social experience" brought about by contemporary technoscientific development.

It was not the intention of this work to arrive at a definitive vision about the problems involved in these different approaches, and, in any case, this would not even be possible, since the questions they pose to us remain open to sociological and political debate. But one thing is certain: We do not consider such approaches to be opposing perspectives, as other works in the social ssciences have suggested. If we were to endorse such a reading, we would turn a blind eye to their obvious commonalities and the contributions they can make to one another.

The relationship we have tried to establish between these different approaches throughout this book allows for a rethinking of environmental sociology that no single perspective or author seems to offer. However, delineating the scope of an environmental sociology here would have been practically impossible, given the wide range of issues it involves. Above all, we hope to have made it clear that an environmental sociology must incorporate, to a greater or lesser extent, the interests and objectives, both sociological and political, of EM, SD, and the theory of RS. This seems possible to us since these concepts and approaches are much closer to each other than has been acknowledged so far.

References

Alexander, Jefrey. 1987. "The Centrality of the Classics." In *Social theory today*, edited by Anthony Giddens and Jonathan H. Turner, 11–57. Stanford, CA: Stanford Univerisity Press.

Allen, John. 1992. "Post-Industrialism and Post-Fordism." In *Modernity and its futures*, edited by Stuart Hall, David Held, and Tony McGrew, 169–204. Cambridge, UK: Polity Press.

Andersen, Skou, and Ilmo Massa. 2000. "Ecological Modernization—Origins, Dilemmas and Future Directions." *Journal of Environmental Policy and Planning* 2, no. 4: 337–45.

Attfield, Robin. 1994. "The Precautionary Principle and Moral Values." In *Interpreting the Precautionary Principle*, edited by Timothy O'Riordan and James Cameron, 152–64. London: Earthscan.

Badham, Richard J. 1986. *Theories of Industrial Society*. New York: St. Martin's Press.

Baker, Susan, Maria Kousis, Dick Richardson, and Stephen Young. 1997. "Introduction: The Theory and Practice of Sustainable Development." In *The Politics of Sustainable Development: Theory, Policy and Practice within the European Union*, edited by Susan Baker, Maria Kousis, Dick Richardson, and Stephen Young, 1–38. London: Routledge.

Barry, John. 1994a. "Discursive Sustainability: The State (and Citizen) of Green Political Theory." In *Contemporary Political Studies*, edited by Patrick Dunleavy and Jefrey Stanyer. Belfast, Northern Ireland: Political Studies Association.

———. 1994b. "The Limits of the Shallow and the Deep: Green Politics, Philosophy and Praxis." *Environmental Politics* 3, no. 3: 369–94.

———. 1995. "Deep Ecology, Socialism and Human Being in the World: A Part of Yet Apart from Nature." *Capitalism, Nature, Socialism* 6, no. 3: 30–38.

———. 1996. "Sustainability, Political Judgement and Citizenship: Connecting Green Politics and Democracy." In *Democracy & Green Political Thought*, edited by Brian Doherty and Marius de Geus, 113–29. London: Routledge.

———. 1999a. *Environment and Social Theory*. London: Routledge.

———. 1999b. *Rethinking Green Politics*. London: Sage.

Barry, John, and Marcel Wissenburg. 2001. "Introduction." In *Sustaining Liberal Democracy. Ecological Challenges and Opportunities*, edited by John Barry and Marcel Wissenburg, 1–15. New York: Palgrave.

Beck, Ulrich. 1987. "The Anthropological Shock: Chernobyl and the Contours of Risk Society." *Berkeley Journal of Sociology* 32: 153–65.

———. 1992. *Risk Society: Towards a New Modernity*. London: Sage.

———. 1995a. *Ecological Politics in an Age of Risk*. Cambridge, UK: Polity Press.

———. 1995b. *Ecological Enlightenment: Essays on the Politics of the Risk Society*. Atlantic Highlands, New Jersey: Humanities Press.

———. 1996. "World Risk as Cosmopolitan Society? Ecological Question in a Framework of Manufactured Uncertainties." *Theory, Culture & Society* 13, no. 4 (November 1996): 1–32.

———. 1997. *The Reinvention of Politics: Rethinking Modernity in the Global Social Order*. Cambridge, UK: Polity Press.

———. 1999. *World Risk Society*. Cambridge, UK: Polity Press.

———. 2000. *What Is Globalization?* Cambridge, UK: Polity Press.

Beck, Ulrich, Anthony Giddens, and Scott Lash. 1994. *Reflexive Modernization: Politics, Tradition and Aesthetics in the Modern Social Order*. Stanford, CA: Stanford University Press.

Bell, Daniel. 1999. *The Coming of Post-Industrial Society: A Venture in Social Forecasting*. New York: Basic Books.

Bennet, John W. 1996. *Human Ecology as Human Behavior*. New Brunswick, NJ: Transaction.

Benton, Ted. 1991. "Biology and Social Science: Why the Return of the Repressed Should be Given a (Cautious) Welcome." *Sociology* 25, no. 1 (February): 1–29.

———. 1999. "Radical Politics—Neither Left nor Right?" In *Theorising Modernity. Reflexivity, Environment and Identity in Giddens Social Theory*, edited by Martin O'Brien, Sue Penna, and Colin Hay, 39–64. London: Longman.

Benton, Ted, and Michael Redclift. 1994. "Introduction." In *Social Theory and the Global Environment*, edited by Ted Benton and Michael Redclift, 1–27. London: Routledge.

Blowers, Andrew. 1997. "Environmental Policy: Ecological Modernization or the Risk Society?" *Urban-Studies* 34 (May): 845–71.

Blühdorn, Ingolfur. 2000. "Ecological Modernization and Post-Ecologist Politics." In *Environment and Global Modernity*, edited by Gert Spaargaren, Arthur P. J. Mol, and Frederick H. Buttel, 209–25. London: Sage.

Bodansky, Daniel. 1994. "The Precautionary Principle in US Environmental Law." In *Interpreting the Precautionary Principle*, edited by Tim O'Riordan and James Cameron, 203–28. London: Earthscan.

Boehmer-Christiansen, Sonja. 1994. "The Precautionary Principle in Germany— Enabling Government." In *Interpreting the Precautionary Principle*, edited by Tim O'Riordan and James Cameron, 31–60. London: Earthscan.

Boland, Joseph. 1994. "Ecological Modernization," *Capitalism, Nature, Socialism* 95, no. 3 (September): 135–41.

Boswell, Christina. 2009. *The Political Uses of Expert Knowledge*. New York: Cambridge University Press.

Boyer, Robert. 1995. *Regulation Theory: The State of the Art*. London: Routledge.

Brown, D. H. 1989. *Social Science as Civic Discourse: Essays on the Invention, Legitimation, and Uses of Social Theory*. Chicago: University of Chicago Press.

Bryant, Christopher G. A. 1995. *Practical Sociology: Post-Empiricism and the Reconstruction of Theory and Application*. Cambridge, UK: Polity Press.

Bryant, Raymond L., and Sinéad Bailey. 1997. *Third World Political Ecology*. New York: Routledge.

Bryant, Christopher G. A., and David Jary. 1991. "Introduction: Coming to Terms with Anthony Giddens." In *Giddens' Theory of Structuration: A Critical Appreciation*, edited by Christopher G. A. Bryant and David Jary, 1–31. London: Routledge.

Buttel, Frederick H. 1987. "New Directions in Environmental Sociology." *Annual Review of Sociology* 13: 465–88.

———. 1996. "Environmental and Resource Sociology: Theoretical Issues and Opportunities for Synthesis." *Rural Sociology* 61, no. 1 (March): 56–76.

———. 1997. "The Politics of Environmental Discourse: ecological modernization and the Policy Process." *Social Forces* 75, no. 3 (March): 1138–40.

———. 2000a. "Classical Theory and Contemporary Environmental Sociology: Some Reflections on the Antecedents and Prospects for Reflexive Modernization Theories in the Study of Environment and Society." In *Environment and Global Modernity*, edited by Gert Spaargaren, Arthur P. J. Mol, and Frederick H. Buttel, 17–36. London: Sage.

———. 2000b. "Ecological Modernization as Social Theory." *Geoforum* 31, no. 1: 57–65.

Buttel, Frederick H., and Peter Taylor. 1994. "Environmental Sociology and Global Environment Change: A Critical Assessment." In *Social Theory and the Global Environment*, edited by Michael Redclift and Ted Benton, 228–55. London: Routledge.

Catton, William R. Jr., and Riley E. Dunlap. 1978. "Environmental Sociology: A New Paradigm." *American Sociologist* 5: 41–49.

———. 1979. "Environmental Sociology." *Annual Review Sociology* 5 (August): 243–73.

———. 1980. "A New Ecological Paradigm for a Post-Exhuberant Sociology." *American Behavioral Scientist* 24, no. 1 (September): 15–47.

Christoff, Peter. 2010. "Ecological Modernisation, Ecological Modernities." In *The Ecological Modernisation Reader. Environmental Reform in Theory and Practice*, edited by Arthur P. J. Mol, David A. Sonnenfeld, and Gert Spaargaren, 101–22. New York: Routledge.

Cohen, Ira. J. 1989. *Structuration Theory. Anthony Giddens and the Constitution of Social Life*. London: Macmillan.

Cohen, Maurie J. 1997. "Risk Society and Ecological Modernisation. Alternatives Visions for Post-Industrial Nations." *Futures* 29, no. 2: 105–19.

———. 1998. "Science and the Environment: Assessing Cultural Capacity for Ecological Modernization." *Public Understanding of Science* 7, no. 2: 149–67.

———. 2000a. "Environmental Sociology, Social Theory, and Risk: An Introductory Discussion." In *Risk in the Modern Age. Social Theory, Science and Environmental Decision-Making*, edited by Maurie J. Cohen, 3–31. London: Macmillan.

———. 2000b. "Ecological Modernisation, Environmental Knowledge and National Character: A Preliminary Analysis of the Netherlands." In *Ecological Modernisation Around the World. Perspectives and Critical Debates*, edited by Arthur P. J. Mol and David A. Sonnenfeld, 77–106. London: Fran Cass.

Collins, Randal. 1999. "Socially Unrecognized Cumulation." *American Sociologist* 30, no. 2 (summer): 41–61.

Connolly, William E. 1983. *The Terms of Political Discourse*. Princeton, NJ: Princeton University Press.

———. 1992. "The Idea of Environment." In *The Environment in Question. Ethics and Global Issues*, edited by David E. Cooper and Joy A. Palmer, 163–78. London: Routledge.

———. 1995. "Other Species and Moral Reason." In *Just Environments: Intergenerational, International and Interspecies Issues*, edited by David Cooper and Joy Palmer, 138–49. London: Routledge.

Crook, Sthepen, Jan Pajulski, and Malcolm Waters. 1992. *Postmodernization: Change in Advanced Society*. London: Sage.

Delanty, Gerard. 1987. *Social Science: Beyond Constructivism and Realism*. Minneapolis: University of Minnesota Press.

Dickens, Peter. 1992. *Society and Nature: Towards a Green Social Theory*. Philadelphia: Temple University Press.

———. 1996. *Reconstructing Nature: Alienation, Emancipation and the Division of Labour*. London: Routledge.

———. 1999. "Life Politics, the Environment and the Limits of Sociology." In *Theorising Modernity. Reflexivity, Environment and Identity in Giddens' Social Theory*, edited by Martin O'Brien, Sue Penna, and Colin Hay, 98–120. London: Longman.

Dobson, Andrew. 1990. *Green Political Thought*. London: Unwin Hyman.

———. 1996. "Democratising Green Theory: Preconditions and Principles." In *Democracy and Green Political Thought. Sustainability, Rights and Citizenship*, edited by Brian Doherty and Marius de Geus, 132–48. London: Routledge.

———. 1998. *Justice and the Environment: Conceptions of Environmental Sustainability and Theories of Distributive Justice*. New York: Oxford University Press.

Doherty, Brian, and Marius de Geus. 1996. "Introduction." In *Democracy and Green Political Thought*, edited by Brian Doherty and Marius de Geus, 1–16. London: Routledge.

Doyal, Len, and Ian Gough. 1991. *A Theory of Human Need*. London: Macmillan.

Dryzek, John S. 1987. *Rational Ecology: Environment and Political Economy*. Cambridge, UK: Blackwell.

———. 1990. *Discursive Democracy: Politics, Policy and Political Science*. Cambridge, UK: Cambridge University Press.

_____. 1997. *The Politics of the Earth: Environmental Discourses.* Oxford: Oxford University Press.

Durkheim, Émile. 1984. *The Division of Labour in Society.* London: Macmillan.

Eckersley, Robin. 1995. "Markets, the State and the Environment: An Overview." In *Markets, The State and the Environment. Towards Integration,* edited by Robyn Eckersley. London: Macmillan.

———. 1996. "Greening Liberal Democracy: The Rights Discourse Revisited." In *Democracy and Green Political Thought. Sustainability, Rights and Citizenship,* edited by Brian Doherty and Marius de Geus, 207–29. London: Routledge.

———. 2000. "Disciplining the Market, Calling in the State: The Politics of Economy-Environment Integration." In *The Emergence of Ecological Modernisation: Integrating the Environment and the Economy?* edited by Stephen C. Young, 233–52. London: Routledge.

Eder, Klaus. 1996a. "The Institutionalisation of Environmentalism: Ecological Discourse and the Second Transformation of the Public Sphere." In *Risk, Environment & Modernity: Towards a New Ecology,* edited by Scott Lash, Bronislaw Szerszynski, and Brian Wynne, 203–23. London: Sage.

———. 1996b. *The Social Construction of Nature: A Sociology of Ecological Enlightenment.* London: Sage.

Ekeli, Kristian S. 1999. *Ethical Perspectives on Sustainable Production and Consumption.* ProSus Working Paper 1/99. Centre for Development an Environment. University of Oslo. www.prosus.uio.no

Ekins, Paul, and Michael Jacobs. 1995. "Environmental Sustainability and the Growth of GDP: Conditions for Compatibility." In *The North the South and the Environment: Ecological Constraints and the Global Economy,* edited by Vinit Bhaskar and Andrew Glyn. Tokyo: United Nations University Press.

Feagin, Joe. 2001. "Social Justice and Sociology: Agendas for the Twenty-First Century." *American Sociological Review* 66 (February): 1–20.

Feagin, Joe, and Hernán Vera. 2001. *Liberation Sociology.* Boulder, Colorado: Westview.

Follesdal, Andreas. 1999. "Sustainable Development, State Sovereignty and International Justice." In *Towards Sustainable Development: on the Goals of Development– and the Conditions of Sustainability,* edited by William M. Lafferty and Oluf Langhelle, 70–83. London: Macmillan.

Giddens, Anthony. 1982. *Sociology: A Brief but Critical Introduction.* London: Macmillan Education.

———. 1984. *The Constitution of Society. Outline of the Theory of Structuration.* Cambridge, UK: Polity Press.

———. 1985. *The Nation-State and Violence.* Berkeley: University of California Press.

———. 1989. "A Reply to My Critics." In *Social Theory of Modern Societies: Anthony Giddens and His Critics.* Cambridge, UK: Cambridge University Press.

———. 1990. *The Consequences of Modernity.* Cambridge, UK: Polity Press.

———. 1991. *Modernity and Self-Identity. Self and Society in the Late Modern Age.* Stanford, CA: Stanford University Press.

———. 1994a. *Beyond Left and Right. The Future of Radical Politics*. Cambridge, UK: Polity Press.

———. 1994b. "Living in a Post-Traditional Society." In *Reflexive Modernization. Politics, Tradition and Aesthetics in the Modern Social Order*, edited by Ulrich Beck, Anthony Giddens, and Scott Lash, 56–109. Stanford, CA: Stanford University Press.

———. 1995. *A Contemporary Critique of Historical Materialism*. Stanford, CA: Stanford University Press.

———. 1996. "Functionalism: après la lutte." In *In Defence of Sociology. Essays, Interpretatons & Rejoinders*, edited by Anthony Giddens, 78–111. Cambridge, UK: Polity Press.

———. 1998a. "The Politics of Risk Society." In *Conversations with Anthony Giddens: Making Sense of Modernity*, edited by Anthony Giddens and Christopher Pierson, 204–17. Cambridge, UK: Polity Press.

———. 1998b. *The Third Way: The Renewal of Social Democracy*. Cambridge, UK: Polity Press.

_____. 2000. *The Third Way and its Critics*. Cambridge, UK: Polity Press.

_____. 2001. "Introduction." In *The Global Third Way Debate*, edited by Anthony Giddens, 1–21. Cambridge, UK: Polity Press.

———. 2009. *The Politics of Climate Change*. Cambridge, UK: Polity Press.

Goldblatt, David. 1996. *Social Theory and the Environment*. Cambridge, UK: Polity Press.

Goodin, Robert. E. 1992. *Green Political Theory*. Cambridge, UK: Polity Press.

Gouldson, Andrew, and Joseph Murphy. 1997. "Ecological Modernisation: Restructuring Industrial Economies." In *Greening the Millennium? The New Politics of the Environment*, edited by Michael Jacobs, 74–86. Oxford: Blackwell.

———. 1998. *Regulatory Realities: The Implementation and Impact of Industrial Environmental Regulation*. London: Earthscan.

———. 2000. "Environmental Policy and Industrial Innovation: Integrating Environment and Economy through Ecological Modernization." *Geoforum* 31, no. 1 (February): 22–44.

Gramling, Robert, and Willian Freundeburg. 1996. "Environmental Sociology: Toward a Paradigm for the 21st Century." *Sociological Spectrum* 16 (October): 347–70.

Grove-White, Robin. 1997. "The Environmental 'Valuation' Controversy: Observations on Its Recent History and Significance." In *Valuing nature? Economics, Ethics and Environment*, edited by John Foster, 21–31. London: Routledge.

Hajer, Maarten. 1995. *The Politics of Environmental Discourse: Ecological Modernization and Policy Process*. Oxford, UK: Clarendon Press.

———. 1996. "Ecological Modernization as Cultural Politics." In *Risk, Environment & Modernity: Towards a New Ecology*, edited by Scott Lash, Bronislaw Szerszynski, and Brian Wynne, 246–68. London: Sage.

Haland, Wenche. 1999. "On Needs: A Central Concept in the Bruntland Report." In *Towards Sustainable Development: on the Goals of Development—and the*

Conditions of Sustainability, edited by William M. Lafferty and Oluf Langhelle, 48–69. London: Macmillan.

Hanf, Kenneth. 1994. "The Political Economy of Ecological Modernization: Creating a Regulated Market for Environmental Quality." In *Privatization and Regulatory Change in Europe*, edited by Michael Moran and Tony Prosser, 126–44. Buckingham, UK: Open University Press.

Hannigan, John A. 1995. *Environmental Sociology: A Social Constructionist Perspective*. London: Routledge.

Hayward, Tim. 2001. "Constitutional Environmental Rights and Liberal Democracy." In *Sustaining Liberal Democracy: Ecological Challenges and Opportunities*, edited by John Barry and Marcel Wissenburg, 59–80. New York: Palgrave.

Holland, Alan. 1997. "Substitutability: or, Why Strong Sustainability is Weak and Absurdly Strong Sustainability Is Not Absurd." In *Valuing Nature? Economics, Ethics and Environment*, edited by John Foster, 119–33. London: Routledge.

Huber, Joseph. 2000. "Towards Industrial Ecology: Sustainable Development as a Concept of Ecological Modernization." *Journal of Environmental Policy & Planning* 2, no. 4: 269–85.

———. 2009. "Ecological Modernization: Beyond Scarcity and Bureaucracy." In *The Ecological Modernisation Reader: Environmental Reform in Theory and Practice*, edited by Arthur P. J. Mol, David A. Sonnenfeld, and Gert Spaargaren, 42–55. New York: Routledge.

Hunt, Jane. 1994. "The Social Construction of Precaution." In *Interpreting the Precautionary Principle*, edited by Timothy O'Riordan and James Cameron, 117–31. London: Earthscan.

Irwin, Alan. 2001. *Sociology and the Environment: A Critical Introduction to Society, Nature and Knowledge*. Cambridge, UK: Polity Press.

Jacobs, Michael. 1991. *Green Economy: Environment, Sustainable Development and the Politics of the Future*. London: Pluto Press.

———. 1997. "Environmental Valuation, Deliberative Democracy and Public Decision-Making Institutions." In *Valuing Nature? Economics, Ethics and Environment*, edited by John Foster, 21–31. London: Routledge.

———. 1999a. "Sustainable Development as a Contested Concept." In *Fairness and Futurity: Essays on Environmental Sustainability and Social Justice*, edited by Andrew Dobson, 21–45. Oxford: Oxford University Press.

———. 1999b *Environmental Modernisation*. London: Fabian Society.

———. 2001. "The Environment, Modernity, and the Third Way." In *The Global Third Way Debate*, edited by Anthony Giddens, 317–39. Cambridge: Polity.

Jänicke, Martin. 1990. *State Failure: The Impotence of Politics in Industrial Society*. University Park: Pennsylvania State University Press.

Jänicke, Martin, Harald Mönch, and Manfred Binder. 2000. "Structural Change and Environmental Policy." In *The Emergence of Ecological Modernisation. Integrating the Environment and the Economy?* edited by Stephen C. Young, 133–52. London: Routledge.

Jänicke, Martin, Harald Mönch, Thomas Ranneberg, and Udo E. Simonis. 1989. "Structural Change and Environmental Impact." *Intereconomics* 24, no. 1: 24–35.

Jayaram, Narayana. 2010. "Revisiting the City: The Relevance of Urban Sociology Today." *Economic and Political Weekly* 45, no. 35 (September): 50–57.

Jessop, Bob. 1990. *State Theory: Putting the Capitalist State in its Place*. University Park: Pennsylvania State University Press.

Jokinen, Pekka, Pentti Malaska, and Jari Kaivo-Oja. 1998. "The Environment in an Information Society." *Futures* 30, no. 6: 485–98.

Kaspersen, Lars Bo. 2000. *Anthony Giddens: An Introduction to a Social Theorist*. Oxford: Blackwell.

Kollek, Regine. 1995. "The Limits of Experimental Knowledge: A Feminist Perspective on the Ecological Risks of Genetic Engineering." In *Biopolitics: A Feminist and Ecological Reader on Biotechnology*, edited by Vandana Shiva and Ingunn Moser, 95–111. London: Zed Books.

Kolm, Serge-Christophe. 2000. *Teorias Modernas da Justiça*. São Paulo: Martins Fontes.

Krohn, Wolfang, and Johannes Weyer. 1994. "Real-Life Experiments." *Science and Public Policy* 21, no. 3: 173–83.

Labaras, Nicos. 2001. "Democracy and Sustainability: Aspects of Efficiency, Legitimacy and Integration." In *Sustaining Liberal Democracy: Ecological Challenges and Opportunities*, edited by John Barry and Marcel, 81–100. New York: Palgrave.

Lafferty, William. M., and Oluf Langhelle. 1999. "Sustainable Development as Concept and Norm." In *Towards Sustainable Development: on the Goals of Development—and the Conditions of Sustainability*, edited by William M. Lafferty and Oluf Langhelle, 1–29. London: Macmillan.

Lafferty, William, and James Meadowcroft. 2000a. *Implementing Sustainable Development. Strategies and Initiatives in High Consumption Societies*. New York: Oxford University Press.

———. 2000b. "Concluding Perspectives." *In Implementing sustainable Development. Strategies and Initiatives in High Consumption Societies*, edited by William M. Lafferty and James Meadowcroft, 422–59. New York: Oxford University Press.

Langhelle, Oluf. 1999. "Sustainable Development: Exploring the Ethics of our Common Future." *Internatinal Political Science Review* 20, no. 2: 129–49.

———. 2000. "Why Ecological Modernization and Sustainable Development should not be Conflated." *Journal of Environmental Policy & Planning*, 2, no. 4: 303–22.

———. 2001. "*Sustainable Production and Consumption—from Conceptions of Sustainable Development to Household Strategies for Sustainable Consumption*." Program for Research and Documentation of a Sustainable Society (ProSus) Report 4/01. Available at http://www.prosus.uio.no/english.

Lash, Scott, Bronislaw Szerszynski, and Brian Wynne. 1996. "Introduction: Ecology, Realism and the Social Sciences." In *Risk, Environment & Modernity: Towards a New Ecology*, edited by Scott Lash, Bronislaw Szerszynski, and Brian Wynne, 1–26. London: Sage.

Lash, Scott, and Brian Wynne. 1992. "Introduction." In *Risk Society. Towards a New Modernity*, edited by Ulrich Beck, 1–19. London: Sage.

Lawrence, Denise L., and Setha M. Low. 1990. "The Built Environment and Spatial Form." *Annual Review of Anthropology* 19: 453–505.

Leiss, William, and Christina Chociolko. 1994. *Risk and Responsibility*. London: McGill-Queen's University Press.

Lélé, Sharachchandra M. 1991. "Sustainable Development: A Critical Review." *World Development* 19, no. 6 (June): 607–21.

Leroy, Pieter, and Jan van Tatenhove. 2000. "Political Modernization Theory and Environmental Politics." In *Environment and global modernity*, edited by Gert Spaargaren, Arthur P. J. Mol, and Frederick H. Buttel,187–208. London: Sage.

Lidskog, Rolf, and Claire Waterton. 2016. "Anthropocene—A Cautious Welcome from Environmental Sociology?" *Environmnetal Sociology* 2, no. 4 (July): 395–406.

Luke, Timothy W. 1995. "Sustainable Development as a Power/Knowledge System: The Problem of "Governmentality."" In *Greening Environmental Policy. The Politics of a Sustainable Future*, edited by Frank Fischer and Michael Black, 21–32. London: Paul Chapman.

Lukes, Steven. 1977. *Essays in Social Theory*. London: Macmillan.

Macarvin, Malcolm. 1994. "Precaution, Science and the Sin of Hubris." In *Interpreting the Precautionary Principle*, edited by Tim O'Riordan and James Cameron, 69–101. London: Earthscan.

Macnaghten, Phil and John Urry. 1995. "Towards a Sociology of Nature." *Sociology* 29, no. 2 (May): 203–20.

———. 1998. *Contested Natures*. London: Sage.

Maldonado, Manuel A. 2018 *Antropoceno: La Política en la Era Humana*. Barcelona, Espanã: Editorial Taurus.

Mandel, Gregory N., and James Thuo Gathii. 2006. "Cost-Benefit Analysis Versus the Precautionary Principle: Beyond Cass Sunstein's Laws of Fear." *University of Illinois Law Review* (August): 1037–80.

Matten, Dirk. 2004. "The Impact of the Risk Society Thesis on Environmental Management in a Globalizing Economy—Principles, Proficiency, Perspectives." *Journal of Risk Research* 7, no. 4: 377–98.

Martell, Luke. 1994. *Ecology and Society: An Introduction*. Amherst: University of Massachusetts Press.

McCormick, J. 1992. *Rumo ao Paraíso: a História do Movimento Ambientalista*. Rio de Janeiro: Relume-Dumará.

McKibben, Bill. 1989. *The End of Nature*. New York: Random House.

McLaughlin, Paul. 2021. "Ecological Modernization in Evolutionary Perspective." *Organization & Environment* 25, no. 2: 178–96.

McManus, Phil. 1996. "Contested Terrains: Politics, Stories and Discourses of Sustainability." *Environmental Politics* 5, no. 1: 48–73.

Meadows, Donella et al. 1972. *Limites do Crescimento: Um Relatório para o Projeto do Clube de Roma sobre o Dilema da Sociedade*. São Paulo: Perspectiva.

Moffatt, Sebastian, and Niklaus Kohler. 2008. "Conceptualizing the Built Environment as a Social-Ecological System." *Building Research and Information* 36, no. 3: 248–68.

Mol, Arthur P. J. 1995. *The Refinement of Production: Ecological Modernization and the Chemical Industry*. Utrecht: Van Arkel.

Mol, Arthur P. J., Joris Hogenboom, and Gert Spaargaren. 2001. "Dealing with Environmental Risk in Reflexive Modernity." In *Risk in the Modern Age. Social Theory, Science and Environment*, edited by Maurie J. Cohen, 83–106. London: Macmillan.

Mol, Arthur P. J., Volkmar Lauber, and Duncan Liefferink. 2000. *The Voluntary Approach to Environmental Policy*. New York: Oxford University Press.

Mol, Arthur P. J., and David A. Sonnenfeld. 2000. "Ecological Modernisation Around the World: An Introduction." In *Ecological Modernisation around the World: Perspectives and Critical Debates*, edited by Arthur P. J. Mol and David A. Sonnenfeld, 3–14, London: Frank Cass.

Mol, Arthur P. J., and Gert Spaargaren. 1993. "Environment, Modernity and the Risk-Society: The Apocalyptic Horizon of Environmental Reform." *International Sociology* 8, no. 4 (December): 431–59.

———. 2000. "Ecological Modernisation Theory in Debate: A Review." In *Ecological Modernisation around the World: Perspectives and Critical Debates*, edited by Arthur P. J. Mol and David A. Sonnenfeld, 17–49. London: Frank Cass.

Mol, Arthur A. J., Gert Spaargaren, and Frederick H. Buttel. 2000. "Introduction: Globalization, Modernity and the Environment." In *Environment and Global Modernity*, edited by Gert Spaargaren, Arthur P. J. Mol, and Frederick H. Buttel, 1–15. London: Sage.

Moser, Ingunn. 1995. "Introduction: Mobilizing Critical Communities and Discourses on Modern Biotechnology." In *Biopolitics. A Feminist and Ecological Reader on Biotechnology*, edited by Vandana Shiva and Ingunn Moser, 1–23 London: Zed Books.

Munslow, Barry, and François Ekoko. 1995. "Is Democracy Necessary for Sustainable Development?" *Democratization* 2, no. 2 (September): 158–78.

Murphy, Joseph. 2000. "Ecological Modernisation." *Geoforum* 31, no. 1 (February): 1–8.

Neale, Alan. 1997. "Organising Environmental Self-Regulation. Liberal Governmentality and the Pursuit of Ecological Modernisation in Europe." *Environmental Politics* 6, no. 4 (November): 1–24.

New, Caroline. 1995. "Sociology and the Case for Realism." *Sociological Review* 43, no. 4 (November): 808–27.

Nielsen, François. 1994. "Sociobiology and Sociology." *Annual Review of Sociology* 20: 267–303.

O'Brien, Martin, and Sue Penna. 1997. "European Policy and the Politics of Environmental Governance." *Policy and Politics* 25, no. 2: 185–200.

O'Brien, Martin, Sue Penna, and Colin Hay. 1999. *Theorising Modernity. Reflexivity, Environment & Identity in Giddens' Social Theory*. New York: Routledge.

O'Riordan, Timothy. 1993. "The Politics of Sustainability." In *Sustainable Environmental Economics and Management: Principles and Practice*, edited by R. Kerry Turner, 37–69. London: Belhaven.

O'Riordan, Timothy, and James Cameron. 1994. "The History and Contemporary Significance of the Precautionary Principle." In *Interpreting the Precautionary Principle*, edited by Tim O'Riordan and James Cameron, 12–30. London: Earthscan.

O'Riordan, Timothy, and Andrew Jordan. 1995. "The Precautionary Principle in Contemporary Environmental Politics." *Environmental Values* 4, no. 3 (August): 191–212.

Offe, Claus. 1992. "Bindings, Shackles and Brakes: on Self-Limitation Strategies." In *Cultural-Political Interventions in the Unfinished Project of Enlightenment*, edited by Axel Honneth, Thomas McCarthy, Claus Offe, and Albrecht Wellmer, 63–94. Cambridge, MA: MIT Press.

Paehlke, Robert C. 1989. *Environmentalism and the Future of Progressive politics*. New Haven, CT: Yale University Press.

Pardo, Mercedes. 1998. "Sociología y Medio Ambiente: Estado de la Cuestión." *Revista Internacional de Sociologia* 20, no. 19: 329–67.

Perrow, Charles. 1984. *Normal Accidents: Living with High-Risk Technologies*. New York: Basic Books.

———. 1994. "Accidents in High-Risk Systems." *Technology Studies* 1, no. 1: 1–25.

Poel, Ibo van de. 2017. "Society as a Laboratory to Experiment with new Technologies." In *Embedding New Technologies into Society: A Regulatory, Ethical and Societal Perspective*, edited by Diana M. Bowman, Elen Stokes, and Arie Rip, 61–87. Singapore: Pan Stanford.

Purvis, Trevor, and Alan Hunt. 1993. "Discourse, Ideology, Discourse, Ideology, Discourse, Ideology." *British Journal of Sociology* 44, no. 3 (September): 473–99.

Redclift, Michael. 1987. *Sustainable Development: Exploring the Contradictions*. London: Routledge.

———. 1992. "The Meaning of Sustainable Development." *Geoforum* 23, no. 3: 395–403.

Redclift, Michael, and Graham Woodgate. 1994. "Sociology and the Environment: Discordant Discourse?" In *Social Theory and the Global Environment*, edited by Michael Redclift and Ted Benton, 51–66. London: Routledge.

Richardson, Dick. 1997. "The Politics of Sustainable Development." In *The Politics of Sustainable Development: Theory, Policy and Practice within the European Union*, edited by Susan Baker, Maria Kousis, Dick Richardson, and Stephen Young, 41–57. London: Routledge.

Rifkin, Jeremy. 1999. *O Século da Biotecnologia: a Valorização dos Genes e a Reconstrução do Mundo*. São Paulo: Makron Books.

Rinkevicius, Leonardas. 2000. "The Ideology of Ecological Modernization." In *"Double-Risk" Societies: A Case Study of Lithuanian Environmental Policy*, edited Gert Spaargaren, Arthur P. J. Mol and Frederick H. Buttel, 163–85. London: Sage.

Roe, Emery M. 1995. "Critical Theory, Sustainable Development and Populism." *Telos* 103: 149–62.

Rosa, Eugene A. 2000. "Modern Theories of Society and the Environment." In *Environment and Global Modernity*, edited by Gert Spaargaren, Arthur P. J. Mol, and Frederick H. Buttel, 73–98. London: Sage.

Sachs, Wolfgang. 1993. "Global Ecology and the Shadow of 'Development.'" In *Global Ecology: A New Arena of Political Conflict*, edited by Wolfgang Sachs, 3–21. London: Zed Books.

Sachs, Noah M. 2011. "Rescuing the Strong Precautionary Principle from Its Critics." *University of Illinois Law Review* 2011, no. 4: 1285–338.

Saunders, Peter. 1989. "Space, Urbanism and the Created Environment." In *Social Theory of Modern Societies: Anthony Giddens and His Critics*, edited by David Held and John B. Thompson, 215–34. New York: Cambridge University Press.

Seippel, Ornulf. 2000. "Ecological Modernization as a Theoretical Device: Strengths and Weaknesses." *Journal of Environmental Policy & Planning* 2, no. 4 (October): 287–302.

Shiva, Vandana. 1995. "Epilogue: Beyond Reductionism." In *Biopolitics. A Feminist and Ecological Reader on Biotechnology*, edited by Vandana Shiva and Ingunn Moser, 267–84. London: Zed Books.

Simonis, U. E. 1985. "Preventive Environmental Policy: Prerequisites, Trends and Prospects." *Ekistics* 52, no. 313 (July/August): 369–72.

———. 1989. "Ecological Modernization of Industrial Society: Three Strategic Elements." *International Social Science Journal* 121: 347–61.

Sjoberg, Gideon, Elizabeth Gill, Norma Williams, and Kathryn E. Kuhn. 1995. "Ethics, Human Rights and Sociological Inquiry: Genocide, Politicide and Other Issues of Organizational Power." *American Sociologist* 25, no. 1 (spring): 8–19.

Skirbekk, Gunnar. 1994a. "Introduction." In *The Notion of Sustainability and Its Normative Implications*, edited by Gunnar Skirbekk, 1–5. Oslo: Scandinavian University Press.

———. 1994b. "Ethical Gradualism, Beyond Anthropocentrism and Biocentrism?" In *The Notion of Sustainability and Its Normative Implications*, edited by Gunnar Skirbekk, 79–126. Oslo: Scandinavian University Press.

Smith, Anthony. 1973. *The Concept of Social Change. A Critique of the Functinalist Theory of Social Change*. London: Routledge.

Smith, Graham. 2003. *Deliberative Democracy and the Environment*. London: Routledge.

Spaargaren, Gert. 1987. "Environment and Society: Environmental Sociology in the Netherlands." *Netherlands Journal of Sociology* 23, no. 1: 54–72.

———. 2000. "Ecological Modernization Theory and the Changing Discourse on Environment and Modernity." In *Environment and Global Modernity*, edited by Gert Spaargaren, Arthur P. J. Mol, and Frederick H. Buttel, 41–66, London: Sage.

Spaargaren, Gert, and Arthur J. P. Mol. 1992. "Sociology, Environment, and Modernity: Ecological Modernization as a Theory of Social Change." *Society and Natural Resources* 5, no. 4: 323–44.

Stapleton, Patricia A. 2016. "From Mad Cows to GMOs: The Side Effects of Modernization." *European Journal of Risk Regulation* 7, no. 3 (September): 517–31.

Stehr, Nico. 1982. "Sociological Languages," *Philosophy of the Social Sciences* 12, no. 1 (March): 47–57.

Strydom, Piet. 2000. *Discourse and Knowledge: The Making of Enlightenment Sociology*. Liverpool, UK: Liverpool University Press.

———. 2002. *Risk, Environment and Society*. Philadelphia: Open University Press.

Sunstein, Cass R. 2005. *Laws of Fear: Beyond the Precautionary Principle*. New York: Cambridge University Press.

Tamanes, Ramon. 1985. *Ecologia y Desarrollo: La Polêmica sobre los Límites al Crescimento*. Madrid: Aliança.

Turner, Bryan S. 1993. "Outline of a Theory of Human Rights." *Sociology* 27, no. 3: 489–512.

World Commission on Environment and Development (WCED). 1987. *Our Common Future*. Oxford: Oxford University Press.

Weale, Albert. 1992. *The New Politics of Pollution*. Manchester, UK: Manchester University Press.

———. 1993. "Ecological Modernization and the Integration of European Environmental Policy." In *European Integration and Environmental Policy*, edited by J. D. Liefferink, Philip Lowe, and Arthur P. J. Mol, 196–216. London: Belhaven.

Weinberg, Adam S., David Pellow, and Allan Schnaiberg. 1996. "Sustainable Development as a Sociologically Defensible Concept." *Advances in Human Ecology* 5 (spring): 261–302.

Wynne, Brian. 1994. "Scientific Knowledge and the Global Environment." *In Social Theory and the Global Environment*, edited by Michael Redclift and Ted Benton, 169–89. London: Routledge.

Yearley, Steven. 1988. *Science, Technology, and Social Change*. London: Unwin Hyman.

_____. 1992a. "Green Ambivalence about Science: Legal-Rational Authority and the Legitimation as a Social Movement." *British Journal of Sociology* 43, no. 4 (December): 511–32.

_____. 1992b. "Environmental Challenges." In *Modernity and Its Futures. Understanding Modern Societies*, edited by Stuart Hall, David Held, and Tony McGrew, 117–53. Cambridge: Polity Press.

———. 1995. "The Environmental Challenge to Science Studies." In *Handbook of Science and Technology Studies*, edited by Sheila Jasanoff, Gerald E. Markle, James C. Petersen, and Trevor Pinch, 457–79. London: Sage.

York, Richard, and Eugene A. Rosa. 2003. "Key Challenges to Ecological Modernization Theory." *Organization and Environment* 16, no. 3 (September): 273–88.

Young, Stephen C. 2000. "Introduction: The Origins and Evolving Nature of Ecological Modernization." In *The Emergence of Ecological Modernisation. Integrating the Environment and the Economy?* edited by Stephen C. Young, 1–39. London: Routledge.

Index

About the Author

Cristiano L. Lenzi is a sociologist and professor of environmental sociology at the School of Arts, Sciences, and Humanities at the University of São Paulo. His studies encompass ecological social theory, environmental politics, sustainability, and environmental citizenship. Among his most recent works is *Transgenics in Dispute: The Political Conflicts in the Commercial Release of GMOs* in Brazil (published in Portuguese and English).

www.ingramcontent.com/pod-product-compliance
Lightning Source LLC
Chambersburg PA
CBHW022315280326
41932CB00010B/1112